IN AND ABOUT THE WORLD

SUNY Series in Science, Technology, and Society
Sal Restivo and Jennifer Croissant, Editors

IN AND ABOUT THE WORLD

Philosophical Studies of Science and Technology

HANS RADDER

State University
of New York
Press

Published by
State University of New York Press, Albany

© 1996 State University of New York

All rights reserved

Production by Susan Geraghty
Marketing by Dana Yanulavich

Printed in the United States of America

No part of this book may be used or reproduced
in any manner whatsoever without written permission.
No part of this book may be stored in a retrieval
system or transmitted in any form or by any means
including electronic, electrostatic, magnetic tape,
mechanical, photocopying, recording, or otherwise
without the prior permission in writing of the publisher.

For information, address State University of New York
Press, State University Plaza, Albany, N.Y., 12246

Library of Congress Cataloging-in-Publication Data

Radder, Hans.
    In and about the world : philosophical studies of science and technology / Hans Radder.
    p.   cm. — (SUNY series in science, technology, and society)
    Includes bibliographical references and index.
    ISBN 0-7914-3049-9 (hardcover : alk. paper). — ISBN 0-7914-3050-2 (pbk. : alk. paper)
    1. Science—Philosophy. 2. Technology—Philosophy. I. Title. II. Series.
Q175.R153 1996
501—dc20
                                                             95-46205
                                                                    CIP

10 9 8 7 6 5 4 3 2 1

CONTENTS

Preface ix

Chapter 1 Introduction: Realization and Nonlocality in Science and Technology   1

Chapter 2 Reproduction and Nonlocality in Experimental Science   9

    2.1 Introduction   9
    2.2 The Realization and Description of Reproducible Experiments   11
    2.3 Reproduction in Experimental Practice   20
    2.4 Normativity, Stability, and Nonlocality   27
    2.5 The Experimenters' Regress   29
    2.6 Data versus Phenomena?   36
    2.7 Experimental Science and Its Social Legitimation   38

Chapter 3 Heuristics, Correspondence, and Nonlocality in Theoretical Science   45

    3.1 Introduction: Intertheoretical Correspondence as a Nonlocal Pattern   45
    3.2 The Generalized Correspondence Principle   48
    3.3 The Correspondence Principle and the Rise of Quantum Mechanics, 1913–1925   53
    3.4 Correspondence in Modern Quantum Theory   57
    3.5 Evaluation of the Generalized Correspondence Principle   59
    3.6 Correspondence and Heuristics   64
    3.7 Philosophical Conclusions   67

Chapter 4 Science, Realization, and Reality   73

    4.1 Change and Work   73
    4.2 Meeting the Kuhnian Challenge: A Referentially Realist Epistemology for Experimental Science   75

4.3	Meeting the Bachelardian Challenge: An Ontology of Persistently Real Potentialities and Historically Contingent Realizations	76
4.4	Realizing Types and Ranges of Reproducibility	80
4.5	The Abstraction (plus Interpretation and Realization) of Nonlocals	83
4.6	Between Transcendental Realism and Constructivism	85
4.7	Experimentation versus Observation?	89

Chapter 5   Normative Reflexions on Constructivist Approaches to Science and Technology   93

5.1	Introduction	93
5.2	Normativity in Constructivism	95
5.3	Reflexivity in Constructivism	97
5.4	Locality	101
5.5	Ontological, Epistemological, and Methodological Relativism	106
5.6	The Actor-Network Theory	109
5.7	Conclusion: Analytical, Critical, and Constructive Reflexivity	115

Chapter 6   Experiment, Technology, and the Intrinsic Connection between Knowledge and Power   119

6.1	Introduction	119
6.2	The Production and Maintenance of Closed Systems	120
6.3	The Relation between Experimentation and Technological Production	123
6.4	The Effects of "Effect Thinking"	129
6.5	The Intrinsic Connection between Knowledge and Power	133

Chapter 7   The Appropriate Realization of Technology: The Case of Agricultural Biotechnology   137

7.1	Introduction	137
7.2	The Potentialities and Actualities of Agricultural Biotechnology and Its Ethics	138

7.3	Realizing Technology	143
7.4	Realizing Appropriate Technology	147
7.5	Realizing Appropriate Agricultural Biotechnology	153
7.6	Conclusion	166

Chapter 8 Philosophy: In and about the World   169

8.1	Introduction	169
8.2	Philosophy as Theoretical	170
8.3	Philosoophy as Normative	175
8.4	Philosophy as Reflexive	183

Notes   *189*

References   *205*

Index   *219*

PREFACE

For a number of years I have been working on an extended research project, lying roughly in the area of overlap between philosophy of science and technology, on the one hand, and science and technology studies, on the other. For more or less contingent reasons, substantial parts of the results of this project have been published separately in various journals and edited volumes. With the present book I intend to recover the coherence of the research project. For this purpose, I have newly written integrative introductory and final chapters. The existing papers have been reedited and integrated along the lines set out in the first chapter. Two papers have been combined, and one has been substantially revised. I would like to thank Steve Fuller for his encouraging support of the project.

The provenance of the chapters is as follows.

Chapter 1 has been composed especially for this book.

Chapter 2 is for the most part taken from: H. Radder, 1995, "Experimenting in the Natural Sciences: A Philosophical Approach," in J. Z. Buchwald, ed., *Scientific Practice: Theories and Stories of Doing Physics* (Chicago: University of Chicago Press), 56–86 (© 1995 by The University of Chicago. All rights reserved). I thank the University of Chicago Press for permission to republish this paper. Marta Fehér, at the Technical University of Budapest, posed the stimulating question of the relationship between materially realizing and witnessing science. Brian Baigrie and other participants in the Conference on "Table-Top Experiments" in Toronto, March 1990, made a number of valuable comments on an earlier draft. I have also profited from a fruitful discussion with my colleagues in our research group on knowledge and normativity at the Vrije Universiteit in Amsterdam. As a new section, I have included in this chapter the main body of H. Radder, 1992, "Experimental Reproducibility and the Experimenters' Regress," in D. Hull, M. Forbes, and K. Okruhlik, eds., *PSA 1992*, Vol. I (East Lansing: Philosophy of Science Association), 63–73. I acknowledge permission by the Philosophy of Science Association to use this material in the present publication.

Chapter 3 has been adapted from H. Radder, 1991, "Heuristics and the Generalized Correspondence Principle," *British Journal for the Philosophy of Science*, 42, 195–226.

The provenance of chapter 4 is H. Radder, 1993, "Science, Realization and Reality: The Fundamental Issues," *Studies in History and Philosophy of Science*, 24, 327–349. I would like to thank my colleague Peter Kirschenmann and two referees of this journal for their helpful comments on an earlier version of this paper.

Chapter 5 has been published as H. Radder, 1992, "Normative Reflexions on Constructivist Approaches to Science and Technology," *Social Studies of Science*, 22, 141–173. An earlier version of this paper was presented at the Joint 4S/EASST Conference, Amsterdam, 16–19 November 1988. I have gained much from commentary on that version by Annemarie Mol and Henk Bodewitz. Comments by some of the referees of *Social Studies of Science* also proved valuable in revising the paper. Tineke van Putten helped to improve the English. Finally, it is a pleasure to thank Richard Gault for the many stimulating discussions we have had about the more general issues related to the subject of this chapter.

Chapter 6 is a substantially revised version of H. Radder, 1986, "Experiment, Technology and the Intrinsic Connection between Knowledge and Power," *Social Studies of Science*, 16, 663–683. In writing the original paper I have greatly profited from discussions with the members of the *Krisis* work group "Alternatives in Science," Peter Groenewegen, Jozef Keulartz, Chunglin Kwa, Peter van Lieshout, Annemarie Mol, and Pieter Pekelharing. I also would like to thank Arie Rip for his valuable comments on an earlier draft of that paper.

Chapter 7 is a slightly adapted version of J. Bunders and H. Radder, 1995, "The Appropriate Realization of Agricultural Biotechnology," in T. B. Mepham, G. A. Tucker, and J. Wiseman, eds., *Issues in Agricultural Bioethics* (Nottingham: Nottingham University Press), 177–204. I thank my coauthor, Joske Bunders, for the stimulating cooperation and for her permission to reproduce the paper in the present form. I acknowledge Nottingham University Press for permission to republish the paper in this book. Finally, I have also benefited from a discussion on the issues in question with the "Werkgroep Wetenschapsonderzoek" at the University of Groningen.

Chapter 8 has been written especially for this volume.

CHAPTER 1

Introduction: Realization and Nonlocality in Science and Technology

At a general level, the philosophical studies of science and technology collected in this volume are circumscribed by the premise that an irresolvable but resourceful tension exists between two basic intuitions. On the one hand, human beings, necessarily, keep trying to grasp and control reality by interpreting it with the help of language and by working in and on it through action. On the other, reality—including natural, human, and social reality—appears to be essentially contingent, complex, and variable. Consequently, for reasons of principle it transcends any attempt at a permanent grasp or control. More in particular, when we study attempts to understand reality theoretically or to intervene in it experimentally and technologically, we observe the same tension. Also in science and technology, an inevitable process of reduction of contingency, complexity, and variability goes along with the recurrent discovery of the limitations and dissolutions of our reductions.

In the chapters that follow, two notions play a central role in specifying and developing these general intuitions. The first is the notion of *realization*. Science, including its epistemic results, and technology, including its "technical" results, must be realized in the world. Doing science or technology requires active interventions in specific parts of the world. Experimental and technological systems need to be realized materially and socially, while new theoretical claims result from a successfully realized transformation of older claims. A crucial implication of this view is that acquiring knowledge cannot be adequately understood as a disembodied cumulation of truths (taking place in the mind or in the realm of ideas). Instead, the notion of realization introduces a kind of *scarcity* in the knowledge production process: the very realization of a specific item or kind of knowledge will frequently exclude the realization of particular alternatives. Thus, new realizations of science or technology do not meet with a blank world but will be confronted with, and have to overcome, existing—and more or less different or conflicting—structurings of the world. For this reason, realizing science or technology is never normatively neutral but rather a process in which knowledge and power are intrinsically connected.

2   IN AND ABOUT THE WORLD

The notion of realization accounts for the tension between contingency and its attempted reductions in this way: Any particular realization remains dependent on human (material and theoretical) work and thus on the caprices of a contingent history. In other words, in scientific and technological practice, the range of successful realizations will be restricted by the complexity and variability of historical developments. However, as I will argue in detail, successful realizations reduce the contingency by making us realize that we are "in touch" with a reality that is bigger than us. In this specific way, the results of science and technology both depend upon and transcend their local realization contexts. Thus, scientific and technological knowledge is, to use Fuller's apt phrase,[1] at once *in* and *about* the world.

A second central notion is that of *nonlocality*. In the successive chapters, this notion will be developed in two different ways. In looking backward, specific, nonlocal patterns can be discerned in the development of science and technology. These patterns denote such features of science and technology that have a significance that clearly goes beyond their local realization contexts. Hence, even if nonlocal patterns do not apply universally, they may possess a broad historical meaning. For instance, since the rise of experimental science, certain notions of reproducibility have been employed in a wide range of practices; or, to give a contemporary example, just a few multinationals have played a worldwide and dominant role in the realization of agricultural biotechnologies. Next, in looking forward all people (including scientists, technologists, and philosophers) reckon with the fact that such nonlocal patterns might, to a greater or lesser extent, continue to shape the future. In this way these extrapolations form the basis of the unavoidable normativity of human agency. An experimenter who sets out to reproduce a particular experiment assumes that it might be reproduced at a different spatiotemporal location and, thus, that its meaning transcends the local context of its original production. Analogously, a critic of the one-sided power relationships in biotechnological developments will be motivated by the assumption that these relationships might well possess a certain (undesirable) persistence.

The notion of nonlocality expresses the tension between contingency and its reductions in different ways. Consider first the nonlocalities discernible by looking backward. On the one hand, these nonlocalities model the relevant practices, processes, and products in a meaningful way. They capture something that is of special significance for our understanding of science and technology. On the other hand, focussing on broader, nonlocal patterns has a prize in that the various details and specificities of the local contexts will be lost out of sight. Consequently, the relevant patterns do *not determine* the scientific or technological practices, processes, and products. Hence, even if we are successful in seeing certain patterns, this does not provide us with a full grasp or control of the patterned situations. Construed in this way, the notion of nonlocal patterns enables us to overcome the inadequate and unfruitful opposition between empirical approaches, which would show us how science and technology "really" work, and philosophical approaches, which would offer only post hoc, rational reconstructions that would not bear upon "real-time" developments in science and technology.

The forward looking, nonlocal extrapolations cannot overcome the tension between contingency and reduction either. Certainly, if we do not wish to proceed blindly, the only alternative is to have us guided by patterns from the past. And surely, if we are successful, they turn out to be a powerful tool in shaping the future. Nevertheless, there is no warrant of successful extrapolation. The future remains contingent, and what was the case up to now may not be so tomorrow. Again, the notion of nonlocality entails that the meaning of scientific and technological practices, processes, and products is neither fully local nor simply universal. This meaning transcends the separate local contexts in a nontrivial way, but its scope is bounded by contingent historical developments.

The merits of the idea of an essential tension between contingency and its attempted reductions must, of course, be shown in the separate chapters of this book. The same applies to the usefulness of the more specific notions of realization and nonlocality. I will conclude this outline of the central notions and claims by briefly sketching the position of the book with respect to four major issues in contemporary debates.

Historical Progress. First, the contingency-reduction idea is not meant to suggest some dialectic of reduction and its negation, followed by a synthesis at a higher level. Nor is a Popperian account in terms of trial and error intended: being confronted with unrealizability and locality should by no means be seen as the result of "making mistakes." Both the dialectical and the Popperian approach entail an untenable notion of historical progress. In contrast, the realization character of science and technology implies not only that various sorts of "progresses" may occur but also that they will be accompanied by various sorts of "losses." They cannot, therefore, be summed up so as to obtain the general historical movement of progress that so many philosophers have been looking for.

(Post)modernism. At the same time, these studies can be seen as an attempt to move beyond the modernism-postmodernism dispute. The limitations and dissolutions of all our attempts at definitely grasping or controlling reality imply the futility of all modernist identities and universalities. Yet, history is not so idiosyncratic as many postmodernists want us to believe. The nonlocalities in the development of science and technology pattern our world in socially and philosophically significant ways. As such, they enable and require theoretical, normative, and reflexive investigations that go beyond the local-universal dichotomy that largely frames and paralyzes the debate between modernists and postmodernists.

Science and Technology. One important nonlocal pattern, which has emerged at least since the second half of the last century, concerns the ever closer relationships between science and technology. Moreover, the resulting "technoscience" has been shaping, and will continue to shape, our material, personal, social, and cultural realities in substantial ways. These developments have undermined the popular contrast between science as aiming at the truth and technology as aiming at socially useful applications. Therefore, the philosophical studies presented in this book do not presuppose strong divides between science and technology, or between epistemic and social issues. More in particular, much attention is given to experimentation, which can be considered as an important link between scientific and technological practice.

Naturalism and Normativism. In recent philosophy of science and epistemology, the claims of traditional, normative approaches have been severely challenged by naturalist critics. Roughly, the former presuppose that we more or less know what science is and that it is the philosopher's task to supply its correct, a priori justification. The latter insist that we still have to find out what science is and that a scientific, instead of a normative, approach is the best or the only way to do so. I agree with the naturalists that philosophers cannot presuppose a sufficient knowledge of what science is. Therefore, many of the analyses and arguments in this book involve or build on detailed studies of the practices, processes, and products of science. However, the naturalist claim that only science can provide such studies will be shown to be both obscure and false. Moreover, if it is not assumed that normative claims should be a priori justifiable, there is no reason at all why philosophical evaluations, critiques, and recommendations should be ruled out in advance.

In arguing for the claims summarized above, this book explores the area of overlap between philosophy of science and technology, on the one hand, and science and technology studies, on the other. In doing so, certain points are particularly emphasized. As philosopher, I stress the role of experimental and technological action and production, and I focus on the connections between epistemic and social issues. As historical and sociological student of science and technology, I argue for the significance of nonlocal patterns, and I point out the importance of the normative dimensions of (the study of) science and technology.

OVERVIEW OF THE BOOK

I will now introduce the issues discussed and summarize the results obtained in the subsequent chapters. Chapter 2 deals with local and nonlocal aspects of experimental practice and experimentally acquired knowledge. First, a philosophical framework for interpreting experimentation in the natural sciences is introduced. Its two main components are the theoretical description and the material realization of experiments. A central issue in this chapter is the question of the reproducibility of experiments. By differentiating between various types and ranges of reproducibility, it is demonstrated that this notion plays a significant role in experimental practice. It functions as a norm that has several nonlocal aspects and that leads to equally differentiated forms of stability of experimental processes and results. Accordingly, experimental practice shows a number of specific interplays of local and nonlocal factors. This general account of experimentation is then employed for the interpretation of more specific issues. First, it is used in an analysis and critical evaluation of the so-called experimenters' regress and of the distinction between (experimental) data and phenomena. In both cases the proposed account of experimentation leads to different and, I think, more adequate views on the issues in question. Next, I deal with the question of the social legitimation of the natural sciences through their utilization in "experimental technologies." The potentialities of materially realizing reproducible and stable experiments prove to be a major component of modern legitimations of the natural sciences. More generally, against the dominant cultural bias towards theoretical and intellectual work, the arguments of this chapter entail a revaluation of the (often neglected) craft work in the natural sciences.

Next, chapter 3 turns to theoretical science. It presents—by means of a discussion and evaluation of the "generalized correspondence principle"—a detailed analysis of the tension between continuity and nonlocality on the one hand, and change and locality on the other. Several philosophers of science have claimed that the correspondence principle can be generalized from quantum physics to all of (physical) science and that in fact it constitutes one of the major heuristic rules for the construction of new theories. In order to evaluate these claims, first the use of the correspondence principle in (the genesis of) quantum mechanics is examined in detail. It is concluded from this and from other examples in the history of science that the principle should be qualified with respect to its nature and relativized with respect to its scope of application. At the same time this conclusion implies a qualification and a relativization of the heuristic power of the principle. Generally speaking, nonlocal intertheoretical correspondences are primarily of a formal-mathematical and experimental but not of a conceptual nature. Moreover, they do not apply to the theories as a whole but only to certain of their parts. In the concluding section, a number of philosophical justifications of the generalized correspondence principle are discussed, and some consequences are pointed out concerning the debates on theory reduction and on the discovery-justification distinction.

Chapter 4 examines the tension between the contingency, complexity, and variability in our reproductions of experimental or observational processes and results, and the human-independent reality to which the knowledge produced by the empirical sciences presumably refers. More in particular, the question to be answered is how the correct claim that laboratory processes and results are produced by human work can be compatible with the equally correct intuition that reality is bigger than us. In developing an adequate answer to this question, the focus is on the ontological aspects of the problem. It is argued that the reproducibility of experiments and observations entails an ontology of both independently persisting, real potentialities and their historically contingent, local realizations. Objective discovery and human production prove to be intrinsically connected. Thus, scientific knowledge can be seen as both in and about the world. Finally, this interpretation is developed by showing how it relates to and differs from transcendental realist and constructivist alternatives, and how it sheds new light on the old debates about the status of universals and about the methodological role of abstraction.

At issue in chapter 5 is the fundamental question of how to reconcile the unavoidable normative commitments in doing research with the apparent relativism that is implied by an equally unavoidable reflexivity. This question is discussed by means of a critical analysis of constructivist accounts of science and technology. Normative reflexivity—in the sense of a systematic and fundamental examination of normative or normatively relevant questions concerning one's own research—is virtually absent in the highly influential constructivist approaches. Therefore, after reviewing and assessing the role of normativity and reflexivity in as far as it has been acknowledged, a number of "normative reflexions" on constructivism are presented. Three general approaches, social constructivism, ethnography, and actor-network theory, are examined in detail. It is shown that constructivist views have a number of normatively questionable consequences. These consequences can be avoided if certain constructivist assumptions—which are empirically unwarranted or unnecessary anyway—are rejected. Instead, a number of alternative assumptions are proposed that suggest an approach to the study of science and technology that is not only empirically adequate but also normatively satisfactory.

The cognitive and normative ambivalences between control and failure of control in experimental and technological contexts are the subject of chapter 6. It starts with a theoretical analysis of the production and maintenance of closed (experimental or technological) systems. On this basis a number of similarities and dissimilarities between experimentation and technological production are discussed. While both imply an active intervention in nature and society, their cognitive and social characteristics may, and often do, display considerable differences. These theoretical conclusions are supported by cases in the areas of nuclear energy and entomology. Next, it is shown that, from the perspective put forward in this chapter, the technology conception prevailing in public debates on technology policy is inadequate; here, the case of nuclear energy serves as the prime example. Finally, it is argued that scientific or technological knowledge and social power are intrinsically connected. From this, a number of conclusions are drawn that may be helpful in arriving at a more adequate appreciation and assessment of the role of science and technology in contemporary society.

Chapter 7 carries these "constructive reflexions" further by proposing a framework for doing normatively relevant research with respect to technology. The framework—in which the notions of realization and non-locality play an important role—consists of two parts. The first part enables an analysis of (existing or intended) technologies in terms of a number of "key features"; the second part proposes a new and generally applicable notion of "appropriate technology." The framework is briefly situated with respect to alternative approaches in technology assessment and, especially, ethics. It aims to be both theoretically adequate in analyzing normative issues in a comprehensive manner and practically applicable in assessing particular technologies and in guiding our actions. In view of this, the theoretical account is applied and specified for the case of agricultural biotechnologies. More in particular, the focus is on problems and opportunities of appropriate technologies for small-scale farmers in developing countries.

Chapter 8 is metaphilosophical in character. The view of philosophy (of science and technology) that is implicit, or only briefly hinted at, in the preceding studies is made explicit and further developed. Philosophy emerges as an activity that is primarily theoretical (either explanatory or interpretative), normative, and reflexive. This conception of philosophy is discussed in relation to other recent approaches. The theoretical aspect is contrasted to the (unduly empiricist) case study approach that dominates a great deal of the historical and sociological studies of science and technology. Along the normative axis, the proposed approach is shown to differ significantly from both cryptonormativist naturalist accounts and untenably universalist views from traditional epistemology. Finally, the reflexive philosophy as advocated here is clearly distinguished from any foundationalism and deconstructionism. Because it is firmly embedded in the world, it does not and cannot claim to possess an Archimedean point upon which to ground its results. But at the same time (and for the same reason!) a full deconstruction of the aboutness of philosophy's claims is no less illusory. In other words, just like science and technology, philosophy is bound to be both in and about the world.

CHAPTER 2

Reproduction and Nonlocality in Experimental Science

2.1 INTRODUCTION

Recent years have seen an increasing wave of studies on experimentation in, especially, the natural sciences. By now a lot of materials and insights have become available concerning this often ignored but crucial aspect of scientific research. In this chapter, I will first describe a philosophical framework for analyzing the practice, process, and products of experimentation in the natural sciences. Then I will discuss certain critical philosophical issues on the basis of this framework. In doing so, I will make use of and evaluate a number of studies from the recent experimental wave, not only philosophical but also historical and sociological ones.

The reference to "philosophical studies" in the subtitle of this book implies a somewhat different perspective than we find in historical and sociological studies, which have constituted the larger part of the experimental wave up to now. In my view, philosophy of science should first of all contribute to a clarification and a critical discussion of the position and function of science in current society, including the intimate connections between science and technology. This, of course, requires a thorough understanding of how science is practiced; and for this purpose, detailed historical and sociological case studies are indispensable.

Yet, the philosopher's aim cannot be an exact and complete description of all phases and aspects of scientific practice. There are, I think, three good reasons for this. The first is that the one and only true account of science does not exist. Scientific practice is a rich and complex activity that can be fruitfully studied from different perspectives at once.[1] In the second place, philosophy approaches science from the angle of a number of distinctively philosophical, theoretical issues, such as realism, rationality, and (social) critique. Issues like these are usually not central to the empirical study of science. Finally, philosophy of science has, in my view, also an important normative component. It does not only aim at understanding but also at assessing science and science-based technology. The most interesting of such normative assessments are those in which the intrinsic connections between epistemic and social aspects are explicitly taken into account. Given my present focus on experimentation, I must postpone a systematic discussion of these metaphilosophical issues until chapter 8. However, at several places the theoretical and normative aspects of the proposed philosophical account of natural scientific experimentation will be illustrated with examples.

A central topic of this chapter is the reproducibility of experiments. As we will see, reproducibility is a rich and complex notion. By differentiating between various types and ranges of reproducibility I will show that, contrary to what is claimed sometimes, reproduction does play a significant role in experimental practice. In addition, reproducibility also functions as a nonlocal norm, the application of which leads to experiments or experimental results that are stable against a number of variations in their local contexts. Therefore, experimental natural science cannot be adequately understood as a mere sum of "local cultures." The general account of experimentation resulting from the discussions in this chapter proves to have significant implications for the interpretation of more specific issues. Below, I will discuss three such issues: the so-called experimenters' regress, the claimed distinction between data and phenomena, and the social legitimation of experimental natural science.[2]

Accordingly, the plan of the chapter is as follows. In section 2.2 I give an analysis of experimentation in terms of the notions of material realization and theoretical description or interpretation. I distinguish three types and four ranges of reproducibility, whose role in experimental practice is examined in section 2.3. Section 2.4 discusses reproducibility as a nonlocal norm and investigates the stability of experiments and their results. Thus, sections 2.2 and 2.3 contain the body of the account of experimentation, while a number of philosophically relevant, general conclusions are drawn in section 2.4. In the remainder of the chapter, these results are developed in three more specific directions. The meaning and implications of the occurrence of an "experimenters' regress" is analyzed, discussed, and evaluated in section 2.5. Generally speaking, my conclusion is that the current interpretation of the regress is not so much false but rather incomplete or one-sided and hence inadequate. In section 2.6 the analysis of experimentation and reproducibility is compared to a recent account of the distinction between "data" and "phenomena." The discussion aims to shed new light on this distinction and to argue for a certain reinterpretation of its meaning. Finally, in section 2.7, I enter into some aspects of the social legitimation of experimental science by means of a comparison of the notions of "witnessing" and "materially realizing" experiments and experimental technologies. Two related views on the issues in question, notably those of Borgmann and Latour, are briefly discussed and evaluated. I conclude the chapter by arguing for a philosophical and social revaluation of the role of craftwork in experimental science and technology.

2.2 THE REALIZATION AND DESCRIPTION OF REPRODUCIBLE EXPERIMENTS[3]

In an experimental process we deal with one or more objects to be studied and with a number of apparatuses. Both object and apparatus may be of various kinds.[4] The experimental process involves the *material realization* and the *theoretical description or interpretation* of a number of manipulations of, and their consequences for, the object and the apparatus, which have been brought into mutual interaction. The general idea is that some information about the object can be transferred to the apparatus by means of a suitable *interaction*. That is, the interaction should produce an (ideally complete) correlation between some property of the object and some property of the apparatus. From this it follows that the theoretical descriptions of object and apparatus should also "interact": they need to have at least some area of overlap. The theory of electrons and the theories of the instrumentation used to experiment on electrons cannot be completely different (see Morrison, 1990, 7–8). In the *detection* stage of the experiment, the information can be obtained by measuring or observing the relevant property of the apparatus. A typical feature of the practice of experimentation is that neither object nor apparatus is "simply available." They have to be carefully *prepared* in agreement with the goal and plan of the experiment.

Consider, for example, an experiment for determining the boiling point of a particular liquid. This liquid is our object under study. Our apparatus consists of a heat source, a vessel, a thermometer, and possibly some supplementary equipment. On the basis of our knowledge of the interaction process between thermometer and liquid, we assume that our readings of the thermometer inform us about the temperature of the liquid. Part of the preparation procedure involves making sure that the liquid in question is pure. This is why it may be necessary first to clean the vessel that will contain the liquid.

The Theoretical Description of Experiments

Let me first discuss the theoretical description. Here it is helpful to make a further distinction between the (intended) theoretical *result q* of the experiment and the other theoretical descriptions *p* that enable us to infer this result. Then, $p \Rightarrow q$ is the specific overall description of the experimental process, whereas *q* refers to its outcome. For instance, in our example of the boiling experiment, the result *q* will be the claim that at a certain time, the fluid in question boils at temperature $y°C$. The description *p*, a conjunction of a number of subdescriptions, is already quite complicated, even in this simple experiment. Description *p* involves, among other things, claims about the liquid (e.g., it is pure; its temperature distribution is homogeneous), about the apparatus (e.g., the temperature remains constant at $y°C$, even if we continue to supply heat), about the interaction between object and apparatus (e.g., the thermometer does not have a substantial influence on the temperature of the liquid), and about the interaction between the experimental system—that is, the sum total of object and apparatus—and its external setting (e.g., the air pressure surrounding the experimental system is constant during the measurements).[5] Generally, the theoretical description contains three main components: it refers to the preparation of object and experimental equipment; to the staging of the processes of interaction and detection; and to the screening and control of potential disturbances from the outside, that is, to the closedness of the experimental system.[6]

The following observations may further clarify the meaning of this notion of theoretical description. First, "theory" is taken here not in the sense of a systematic and comprehensive theory, but in a vaguer and more restricted sense. The prime aim of the theoretical description is not to offer a systematic explanation of the phenomenon under study but to supply a "singular causal explanation" of how the result q is produced (cf. Bogen and Woodward, 1988, 322). Theory in this sense includes, in terms of Hacking's taxonomy of the elements of laboratory experiments, background knowledge about object and apparatus, topical hypotheses, and assumptions about data processing (Hacking, 1988a, 508–511). Alternatively, one may think of Pickering's notions of instrumental and phenomenal models (Pickering, 1989). Second, the theoretical description $p \Rightarrow q$ does not give us a complete account (whatever that may be) of the experimental process. However, such an account is not necessary. We need to consider only those aspects of the experimental process that are deemed *relevant* to obtaining the intended result. For instance, in the boiling experiment discussed above, most of the specific characteristics of the heat source as well as a description of the gravitational field in the laboratory are irrelevant to successfully determining the boiling point of the liquid. Third, in experimental practice the argumentation by which we infer the result q from the premises p will not always be explicit. I assume, however, that if asked the experimenter will or should be able to come up with a plausible story about how the experimental result is produced.

The Material Realization of Experiments

Next we need an explication of the notion of the material realization of an experiment (see also Radder, 1988, 69–76). The question is whether there is a way to describe and analyze the performance of an experiment other than in terms of the particular theoretical description that is in fact used to perform it. The main problem can be illustrated with this example. Suppose we want to determine experimentally the mass of an object that is at rest in relation to the measurement equipment. Two scientists each carry out such an experiment in the same way. Nevertheless, one interprets the actions performed as a measurement of the Newtonian mass; and the other, as a determination of the Einsteinian mass. But both performed "the same" actions and thus—in a certain sense—the same experiment. Therefore, if we want to describe experimental action unambiguously, we have to find some sort of abstraction of these various specific theoretical interpretations. Yet, it is an indisputable fact that concrete experimental action is always action on the basis of certain theoretical ideas: without theoretical ideas there can be no experiments.

I propose the following maneuver to make the distinction and the interplay between the theoretical description and the actual performance of experiments more explicit and manageable. Let us suppose that in the example of the experiment to determine the boiling point of a liquid, the experimenter A orders B, a complete layperson in the field of heat theory, to actually carry out all the necessary actions. A tells B for example: "Take this thing (here) and put it in the holder (over there), in such a way that its lower part hangs in the liquid; then watch the rising silver column; when it does not rise anymore for a bit, then write down (here) which point (which number) is reached by the upper end of the column"; and B does all this *correctly according to* A.

The general supposition is that experimenter A can delegate the actual carrying out of experiments to one or more laypersons B, that is, to people who have no knowledge of the theoretical interpretations. I am referring to actions like the installation and operation of various experimental apparatuses; the preparation of the necessary substances and material; the reading of pointer positions on graduated scales, of color changes, of traces on photographic plates, etc. This all happens under A's supervision and with perhaps A's correcting instructions to guarantee that the experimental result is obtained in the right way according to the experimenter. With the help of this maneuver I can now define the concept of the *material realization* of an experiment as the whole of the experimental actions that are carried out by B in a correct way according to A and that can be described in A's instructions to B in the language in which A and B communicate with each other.[7]

This definition of the material realization of an experiment is based on two general characteristics of the human condition. The first is the possibility of a *division of labor*. Not only experimenters but all human beings are themselves part of nature. Therefore, by materially interacting with surrounding nature, all human beings are able to produce certain situations in experimental systems. In the second place, the procedure of material realization exploits the fact that in any particular experiment, some sort of language exists in which a layperson and an experimenter can communicate with each other about certain fundamental aspects of the experimental production process in a usually reliable and successful way. I will call the language thus defined "common language."[8]

So far I have dealt with the case in which experimenter A already knows how to successfully perform the experiment. In this case, the procedure of material realization introduced above comes down to a specific kind of repetition of the experiment. The prime aim of the proposed maneuver is to make visible the significance of experimental action and production. Apart from this it may also, in the form of A's *correcting* instructions, exhibit aspects of the process through which B learns to carry out the experiment. Therefore, if B's performance succeeds, the maneuver also shows us the "learnability" of A's productive skills. But the procedure of material realization through division of labor may also be applied in the more general case in which both A and B still have to find out how to do the experiment and what experimental success means in the case in question. In this case the *inter*action between A's developing theoretical insights and B's growing practical skills will come more to the fore.[9] However, for most purposes of the present chapter, a central topic of which is, after all, the *re*producibility of experiments, it will be sufficient to consider cases of material realization that aim to describe A's previously acquired practical abilities.

What about the role of theory in materially realizing experiments? As I have stressed, without theoretical knowledge, experimenting would be impossible. The instructions and corrections of the experimenter are guided by his or her scientific insights concerning the preparation of object and equipment, the suitable staging of the interaction and detection, and the effective control of disturbing influences. Nevertheless, the resulting description of the material realization is phrased in common language and thus it is, in this specific way, theory independent. As we will see further on, this procedure suggests an *empirical mechanism* by which it is possible to step from one theoretical description of an experiment to another. Put differently, it will enable us to define the reproducibility of the material realization of an experiment under a whole set of, possibly radically different, theoretical interpretations.[10]

In sum, with the term "material realization" I intend to refer to experimental action and production *either* by the experimenters themselves *or* by laypersons. But in order to make explicit these aspects of experimentation and the specific way in which they are theory independent, we need the proposed maneuver that makes use of the possibility of division of labor and of a common language as a means of communication between scientists and laypersons.

Two Illustrations

I will continue the discussion of the material realization of experiments by addressing the question of the practical feasibility of carrying out experiments in the way described by the definition of material realization. For this purpose, consider the following quotation regarding an ethnomethodological study carried out by Schrecker in a completely different context:

> In Schrecker's study, a methodological set-up was used for the purpose of perspicuously identifying the mutual dependence of chemical reasoning and embodied action. Schrecker volunteered his services to aid a handicapped student in his laboratory work for an undergraduate chemistry course. He then received the student's permission to use his experimental work as a research topic. There resulted a division of labour and responsibility between Schrecker, who was largely ignorant of the field of chemistry—and ignorant as well of the specific lab assignments he bodily assisted—and Gordon, the chemistry student, who because of a spinal injury was paralyzed from the neck down with very limited use of his hands. Gordon depended upon Schrecker to bodily perform the work at the bench for the weekly lab experiments assigned to students in the course on "Quantitive Analysis."
>
> The isolation of Schrecker's handiwork from its theoretical basis in chemistry necessitated that Gordon and he make explicit for one another how the experiment was progressing as a witnessable production of chemistry. Gordon provided instructions for Schrecker on what to do next, while at the same time he relied upon Schrecker's developing work to show him what "next" meant in terms of where they stood in the course of the experiment's events.[11]

This case shows directly the practical possibility and the main features of materially realizing experiments by means of a division of labor between an experimenter and a layperson. In this study, just as in the definition of "material realization," we clearly have to do with a somewhat unusual maneuver; we have a layperson, Schrecker, who was "recognizably doing chemistry, whether he knew it or not" (Lynch, Livingston, and Garfinkel, 1983, 228); and we also see the specific interplay between theoretical knowledge and experimental action and production.

A second practical illustration is slightly less direct but nevertheless illuminating. It concerns Boyle's so-called void-in-the-void experiment carried out in the 1650s, which is summarized by Shapin and Schaffer as follows:

> This is what Boyle did: he took a three-foot-long glass tube, one-quarter inch in diameter, filled it with mercury, and inverted it as usual into a dish of mercury, having, as he said, taken care to remove bubbles of air from the substance. The mercury column then subsided to a height of

about 29 inches above the surface of the mercury in the dish below, leaving the Torricellian space at the top. He then pasted a piece of ruled paper at the top of the tube, and, using a number of strings, lowered the apparatus into the receiver. Part of the tube extended above the aperture in the receiver's top, and Boyle carefully filled up the joints with melted diachylon. . . .

Pumping now commenced. The initial suck resulted in an immediate subsidance of the mercury column; subsequent sucks caused further falls. . . . After about a quarter-hour's pumping . . . , the mercury would fall no further. (Shapin and Schaffer, 1985, 42–43)

Although in this summary no direct mention is made of laypersons, Boyle in fact nearly never carried out the experiments himself. He had this done by all sorts of assistants, "who had skill but lacked the qualifications to make knowledge" (Shapin, 1988, 395). Apart from this, the above passage provides a beautiful illustration of how an experiment is materially realized. Shapin and Schaffer, moreover, explicitly distinguish between, on the one hand, the experimental production of "matters of fact" and, on the other, Boyle's theoretical interpretations on the basis of his ideas about the nature and properties of the air, such as its pressure and "spring":

So far, the account we have given has been restricted to what Boyle said was done and observed, without any of the *meanings* he attached to the experiment. (Shapin and Schaffer, 1985, 43)

From such examples it is, I think, justified to conclude that materially realizing an experiment in the way defined above is not merely a philosophical construction but is also feasible and plays a role in actual experimental episodes.

Types and Ranges of Reproducibility

Many philosophers assume, implicitly or explicitly, that successful experiments are or should be reproducible.[12] However, since "experiment" is a general term for what in fact is a rather complex process, the precise meaning of this assumption is not clear. To clarify the notion of reproducibility we need to address the following question: reproducibility *of what* and *by whom*? In answering this question I will first distinguish, on the basis of the above analysis of experimentation, between three types of reproducibility and discuss their mutual relations.

Reproducibility of the Material Realization of an Experiment. This type of reproducibility obtains when the same material realization can be reproduced under different interpretations, such as $p \Rightarrow q$, $p' \Rightarrow q'$, $p' \Rightarrow q$, or $p \Rightarrow q'$. Since in this case a reproduction may be achieved on the basis of any member of a whole class of theoretical interpretations, reproducing the material realization of an experiment does not depend on (a shared belief in) some *particular* interpretation from that class.

Consider, for instance, the experimental determination of the Newtonian mass of an object that does not move relative to the measuring apparatus. Suppose that experimenter A has a setup available and has the experiment performed successively by layperson B and by experimenter A' who, for the time being, acts as a layperson. A instructs both B and A', in common language, to carry out the experiment. Now it may well be that the performance of the experiment requires some special skills that are, for instance, mastered by B but not yet by A'. Then of course the two descriptions of the material realizations resulting from this first run will be different, and another run will be necessary. The assumption is, however, that after some time of training, the two descriptions of the material realizations will converge. Yet, having reached this phase, the theoretically informed experimenter A' may rightly claim to have measured the Einsteinian mass, in contrast to A who claims to have determined the Newtonian mass. In this case we can say that the material realization of the experiment has been reproduced, even though its theoretical description or its result is radically at variance in the two experiments.

The possibility of reproducing material realizations is based on the fact—demonstrated in processes of division of labor—that performing an experiment may be learned by people who are ignorant of, or disagree with, the theoretical interpretation of the (results of the) actions they produce. In such processes of division of labor, communication through common language plays a crucial role. The range of this type of reproducibility is greatly enhanced by the fact that experiments can be made more robust and less dependent on the skills of specific individuals, as happens for instance in their "demonstrative" or "showing" stage.[13] A reproducible material realization indicates that *a* stable phenomenon has been detected. It does not imply any agreement about *what* phenomenon it is. It is even possible that some interpreters will argue that the phenomenon is an artifact, because, though it is stable, it is not to be attributed to the object under study but to certain features of the apparatus.

Reproducibility of an Experiment under a Fixed Theoretical Interpretation $p \Rightarrow q$. We might find out, for example, that in our boiling experiment, the description $p \Rightarrow q$ does not lead to reproducibility if q is the claim that the liquid boils at $y°C$. However, if we add an error interval z and change q into the claim that the boiling point lies within the interval $(y\pm z)°C$, then the experiment may well be reproducible under this (newly) fixed theoretical description. Or, we might find that a description p that does not include the air pressure as a relevant factor does not lead to a reproducible result, whereas inclusion of the air pressure does. Thus, this type of reproducibility implies a repeatability of the experiment from the point of view of the theoretical interpretation in question. If a particular reproduction succeeds, the people involved believe that what has been reproduced is the experimental process as described by this theoretical interpretation. As we shall see further on, in section 2.3, reproducing an experiment under a fixed theoretical interpretation will not always imply exactly reproducing the same material procedures. Sometimes slight differences occur, in which case we have an approximate reproduction of the material realization.

Reproducibility of the Result q of an Experiment. This type of reproducibility applies when it is possible to obtain the same experimental result by means of a set of different experimental processes. Again, "sameness" and "difference" mean sameness and difference from the point of view of the theoretical interpretation. Thus, we may have $p \Rightarrow q$, $p' \Rightarrow q$, or $p'' \Rightarrow q$, where p, p' and p'' are (possibly radically) different descriptions that will usually (but not necessarily) describe different material procedures for realizing q. A simple example is a determination of the boiling point of our liquid by means of different types of thermometers, for instance, a mercury and a gas thermometer. I propose to call an experiment that reproduces q but not p a *replication*.

It follows from the above definitions that the reproducibility of an experiment under a fixed overall description and the replicability of the result do not mutually imply each other. Furthermore, it will be clear that when we reproduce a result, we do not necessarily reproduce the material realization of the first experiment. Though it may be that p' is simply a different interpretation of precisely the same material realization, p' may just as well describe a slightly or even completely different experimental process that can be used to determine the same result q.

Reproducing an experiment, in any of its senses, requires two things. In the first place, it requires agreement on the description of the original experiment. This description provides a first specification of the ways in which original and reproduced experiment should be similar. However, having such a description is not enough. Also required are the abilities and resources for skillfully re-producing the experimental situations described. In spite of these conditions, claims on reproducibility possess a definite surplus value. This value rests on the fact that agreement on descriptions and availability of the relevant skills and resources cannot guarantee success. Thus, obtaining reproducibility constitutes a significant experimental achievement.[14]

Given the three kinds of descriptions of an experiment (the description of its material realization, the theoretical description of the overall experimental process, and the theoretical description of its result), the question "reproducibility of what?" can now be answered as follows. What is reproduced are the experimental situations identified by any of these three kinds of description.[15] Finally, with respect to the question "reproducible by whom?" four ranges suggest themselves: reproducibility by any scientist or even by any human being, in the past, present, or future; reproducibility by contemporary scientists; reproducibility by the original experimenter; and reproducibility by the lay performers of the experiment. Combining the three types with the four ranges of reproducibility results in twelve possibilities, which are summarized in table I.

Nature and Scope of the Approach

So much for the definitions of these different notions of reproducibility. In the next section, I will come back to this issue and scrutinize what role these notions play in actual experimental practice. In concluding this section I want to say something about the nature and scope of the foregoing analysis of the practice, process, and products of experimentation. As stated in the introductory section, the nature of my analysis is philosophical, which implies that it is both theoretical and normative. Here I will deal with the first aspect; a discussion of the normative aspect will be given in section 2.4.

TABLE I
Types and Ranges of Reproducibility

By Whom?	Of What?		
	Reproducibility of the Material Realization	*Reproducibility of the Theoretical Interpretation*	*Reproducibility of the Result of the Experiment*
By any scientist or any human being, in past, present, or future	1	5	9
By contemporary scientists	2	6	10
By the original experimenter	3	7	11
By the lay performers of the experiment	4	8	12

Consider the notion of material realization. It is important to distinguish between the process of material realization itself and its description by means of the theoretical notion of material realization as introduced above. The claim is that material realization, the process of acting and producing, is a fundamental aspect of every experiment. The *notion* of material realization, however, is a theoretical-philosophical concept. Like many other theoretical concepts, its chief function is not to describe but to explain or interpret a fundamental aspect of experimental practice. It purports to explain or interpret both the interaction between and the relative autonomy of experimental action and theoretical reasoning. Put differently, it attempts to explicate the mechanism through which theoretical interpretations are underdetermined, not by data or empirical statements, but by the process of experimental action and production. Conversely, it will be equally clear that theoretical notions should not be mere speculations that are totally unrelated to the practice of science. Therefore, I have also pointed out that the concept of material realization can be empirically substantiated through the phenomenon of division of labor in experimental practice.

What about the scope of the above approach to experimenting? My claim is that it is generally applicable to experiments in the natural sciences.[16] Although the illustrations given so far concern small-scale experiments, this does not entail that the analysis only applies to yesterday's experiments. Contrary to popular opinion, small-scale experiments still occupy a prominent place in contemporary natural science. Think, for instance, of the recent cold fusion and high-temperature superconductivity experiments. Or consider the experiments in fluid dynamics, which played such an important role in the recent development of chaos theories (Gleick, 1987, 125–131 and 191–211). Moreover, the analysis does not seem to be restricted to tabletop experiments either. Of course, in complex, large-scale experiments it will be difficult to carry out the analyses along the above lines in full detail. Yet, as far as I can see, there do not seem to be any fundamental reasons why the approach would not, in principle, be applicable to "big science" experimentation too.

2.3 REPRODUCTION IN EXPERIMENTAL PRACTICE

I will now consider the three types of reproducibility and examine how far they range in actual experimental practice. Of course, there is a difference between the claim that scientific experiments are or should be reproducible and the claim that they, in fact, are or should be reproduced by scientists (see chapter 4). I think that many philosophers, when discussing this issue, intend to refer primarily to the possibility and not to the actual occurrence of reproduction. But without a thorough empirical investigation into the actual occurrence of reproduction, we cannot decide whether the philosophical claim is merely an idle speculation or has a firm footing in experimental practice. At this point, however, we are confronted with the claims of empirical students of science. Collins, for instance, states that "replication of others' findings and results is an activity that is rarely practised" (1985, 19). And Hacking asserts that, "roughly speaking, no one ever repeats an experiment" (1983b, 231).

I do not think that these statements, formulated in this manner, are strictly speaking false. My main objection, though, is that they are inadequate, because they cover only some of the issues surrounding experimental reproduction. By distinguishing between reproduction "of what" and reproduction "by whom" we obtain a richer notion of experimental reproduction, which enables us to draw a more finely grained map of its place in experimental practice (see table I). As I will show now, this more differentiated approach makes it possible to uphold the claim that various types of reproducibility do in fact play a significant role in experimental practice.

Reproducing the Material Realization

Let me start with the first type of reproduction: the reproduction of the material realization of an experiment being compatible with a whole set of theoretical interpretations. This type concerns reproducing an experiment as defined by the actions of laypersons or, as the case may be, of the experimenter. I will first give an example and then point out the problems one may come across in trying to reproduce a material realization.

Consider, for instance, Boyle's experiments with the air pump. As we have seen, Boyle almost never performed the physical manipulations himself. This was done by various sorts of assistants, being for the most part laypersons. According to Shapin, "they made the machines work, but they could not make knowledge" (1988, 395). Therefore, every successful repetition of an air-pump experiment performed in this manner—for instance, a successful trial at Boyle's home and a showing in the rooms of the Royal Society—reproduces its material realization. The material realization remains the same, even if the theoretical interpretations differ. If, for example, we place a burning candle in the receiver of the air pump and notice that the candle goes out after some time of pumping, we have a stable material realization, whether we interpret the ceasing of the flame as due to the vanishing air, as Boyle did, or as due to the violent winds produced by the pumping, as Hobbes did (see Shapin and Schaffer, 1985, 122).

In general, this type of reproducibility applies to the demonstrative stage of experiments and a fortiori to the large number of more or less standardized experiments, such as the measuring of a boiling point. As we have seen, reproducing an experiment under a fixed theoretical interpretation implies the (approximate) reproduction of its material realization. Consequently, the examples of the former type of reproduction given below are also examples of the latter type.

Nothing I have said so far should be taken to imply that this type of reproduction can be achieved easily, let alone in an algorithmic manner. In many cases, craftsmanship is required to reproduce material realizations.[17] If this skillfulness is absent and if it cannot be learned by laypersons or other experimenters, attempts at reproducing the material realization will fail. As Shapin remarks about Boyle:

> Time after time in Boyle's texts, technicians appear as sources of trouble. They are . . . responsible for pumps exploding, materials being impure, glasses not being ground correctly, machines lacking the required integrity. (Shapin, 1988, 395)

In general, there seem to be three kinds of trouble with respect to reproducing a material realization. First, the required craftwork might be so extraordinary that only one or a few people are able to master it. In such a case successful intervention in nature would require a very individual "feeling."[18] In the second place, it may be impossible for all experienced performers, including the original experimenter, to reproduce the required material actions and processes. Reproducing the material realization will fail if one cannot correctly repeat the preparation of object and equipment, the staging of their interaction, and the creation and maintenance of the external conditions required for the closedness of the experimental system. Third, we also need social control of the people involved in the experiment. We must be able to supervise and discipline the experimenters or laypersons in such a way that they will exercise sufficient care in performing their tasks (see Rouse, 1987, esp. 220–226). And we must control any other people who might come into contact with the experimental setting, in order to prevent disturbing influences from the outside. That is, we must also create and maintain the social conditions for the closedness of the experimental system (see the analysis and examples in chapter 6).

Notwithstanding these potential problems, the claim that many, if not most, material realizations in the natural sciences can in fact be reproduced by the original experimenters, contemporary scientists, and lay assistants does seem to be plausible. Moreover, in the case of more or less standardized experiments, the possibilities of reproduction will stretch considerably beyond the period of their original performance.

Replication of the Experimental Result

Next there is what I have called the replication of the result q of an experiment. In most cases this concerns experiments carried out by different experimenters, in which it is possible to determine q by means of a number of different methods. In these cases, therefore, we do not have a reproduction of the same material realization. The scientists' reasons for preferring replication to exact reproduction are recorded by Collins:

> Some said that an exact copy could gain them no prestige. If it confirmed the first researcher's findings, it would do nothing for *them*, but would win the Nobel prize for *him*, while on the other hand, if it disconfirmed the results there would be nothing positive to show for their work. But, if their apparatus was better in some way, in the case of positive results, they would be ahead of the field, and in the case of negative findings they could be seen as being better experimenters than the first researcher. (Collins, 1975, 210)

More in particular, there is a drive to take advantage of local opportunities. As one of Collins's interviewees said:

> So what you copy is the part that you are not an expert on and the part you are an expert on you build something you think is better.[19]

The last quotation shows fittingly what may go wrong in attempts at replication. First, building an exact copy of part of the setup may fail. After all, for most replications the original experimenter will not be on the spot to instruct, supervise, or correct the replicators. Furthermore, published accounts of the original experiments are apt to be schematic and incomplete, so that problems can be expected to arise. Second, a general consensus about what constitutes good equipment for measuring q may (still) be lacking. A fortiori, this implies that there will be no agreement among experimenters on whether or not some part of the equipment is better for measuring the experimental result. So here too problems for replication may arise. Some well-known problems with replication documented in the literature concern attempts at detecting gravity radiation (Collins, 1975), experimental tests of local hidden variable theories (Harvey, 1981), and efforts at replicating a number of Boyle's air-pump results (Shapin and Schaffer, 1985, ch. VI).

Yet even radical relativists concede that in scientific practice sooner or later these controversies fade away and a more or less stable consensus on procedures of replication develops. A paradigm case is the experimental replication of tests of Avogadro's hypothesis, which says that different substances of one mole all contain the same number of molecules. In 1913, the French physicist Perrin listed a large number of different methods for checking Avogadro's hypothesis.[20] By that time it had been tested by a set of quite different experimental procedures, for example in Brownian motion, alpha decay, X-ray diffraction, blackbody radiation, and electrochemical processes. By decreasing the chance of systematic errors and by increasing the systematic nature of Avogadro's hypothesis, these independent replications contributed much to its plausibility. So here is a further reason for preferring replication to strict repetition. If successful, it not only adds to the credibility of the experimenters in question but also enhances the systematic character and hence the plausibility of the experimental result.

Reproductions under a Fixed Theoretical Description

In the case of reproducing an experiment under a fixed theoretical description we are, in contrast to replication, concerned with a strict repetition of the experiment from the point of view of the theoretical characterization. It seems to me that this is the primary sense of reproducibility most philosophers have in mind. Thus, in Dingler's view, it is only the realization of the "completely unambiguous" theoretical concepts of what he calls the "ideal sciences" (arithmetics, chronometrics, geometrics, and mechanics) that makes possible and guarantees the unambiguous reproduction of experiments (see Dingler, 1952, 21). Popper states that "the scientifically significant *physical effect* may be defined as that which can be regularly reproduced by anyone who carries out the appropriate experiment in the way prescribed" (Popper, 1965, 45). And Habermas claims that repeating a (successful) experiment under exactly the same conditions must lead to the same effect (Habermas, 1978, 127). Note that in the context of my analysis of experimentation, it is the theoretical characterization that decides what is meant by "exactly the same." This implies that the actual material circumstances are allowed to vary in a series of reproductions *as long as* these variations are deemed irrelevant and, thus, do not show up in the theoretical description. Below we will meet with examples of such slight variations in the material realization.

Let me *first* consider attempts by different experimenters to exactly reproduce an experiment. Such attempts at what has also been called "isomorphic reproduction" (Collins, 1985, 170–171) are rather exceptional, for reasons explained in the discussion of replication. Yet they do occur. Collins classifies two experiments from his case studies, one concerning gravity radiation and one concerning psychokinesis, as "fairly isomorphic" (Collins, 1985, 170). In general, when reproductions of this type, performed by different scientists, occur, much is at stake in confirming or disconfirming the claimed results; in other words, the claims of the original experiment are both important and controversial.

Second, in experimental practice many reproductions of the type under discussion are carried out by the original experimenters. They do so in order to make sure that their experiment really works. For instance, Shapin and Schaffer report about Boyle's void-in-the-void experiment discussed in section 2.2:

The experiment was quickly repeated in the presence of witnesses, and the same result was obtained.²¹

As a second example, consider the experiments by the Italian physicist Morpurgo, in the 1960s and 1970s. These experiments were aimed at the detection of free, fractional electric charges, that is, free quarks (see Pickering, 1989). At one stage of a series of experimental runs stability broke down: Morpurgo started to find both integral and fractional charges. After some time, however, he discovered how to restore an interactive stability between his material procedures and a new theoretical interpretation of the working of his apparatus. As Pickering reports:

> He increased the separation between the metal plates . . . and found that he could obtain consistent measurements of only integral values in this way. (Pickering, 1989, 288)

Thus, the latter measurements constituted a series of reproductions under a (newly) fixed theoretical interpretation.

Third and most important, this type of reproduction by the same experimenter also plays a role in most experiments in a slightly different way. Consider, again, Boyle's void-in-the-void experiment. Boyle repeated this experiment with a smaller receiver but still observed exactly the same fall of the mercury column. He concluded that the size of the receiver was not a relevant factor in his theoretical interpretation.²² We can render this procedure more generally as follows: any experimenter who repeats an experiment under varied (material) conditions and finds that these conditions do not make a difference will conclude that they are irrelevant to the theoretical interpretation of the experimental process. Therefore, in this way he or she will have reproduced the experiment from the point of view of this theoretical interpretation. Since such a procedure is quite common, this type of reproduction is prominently present in experimental practice.²³

A *fourth* way of reproducing an experiment under a fixed theoretical interpretation also occurs frequently. This is reproducing an experiment as part of another experiment. It happens whenever established experimental procedures or stable, experimentally produced phenomena and entities are employed as parts of a larger experiment that is performed for the first time. Cases like these concern the reproduction of experiments that are more or less standardized or are otherwise deemed uncontroversial. There is agreement about some theoretical description and about how to correctly perform the experiments on the basis of this description. Examples of these cases abound. They range from measuring time by means of a clock as part of an experiment to spraying artificially produced positrons into an experiment for detecting free quarks.²⁴

Of course, in all these cases success is not guaranteed, for reasons that will be clear from the preceding discussion: agreement about the theoretical description may break down and/or problems may arise in trying to (approximately) reproduce the material realization. Yet, I think that the previous analysis shows that reproducing an experiment under a fixed theoretical interpretation does occur frequently in scientific practice, even if attributing reproducibility to a particular experiment is fallible in principle. Moreover, this type of reproduction is not at all trivial or insignificant. In most cases it requires hard and ingenious theoretical and experimental work to achieve.

Conclusion

On the basis of the broader notion of reproducibility summarized in table I, it is possible to maintain the claim that reproduction does play a significant part in actual experimental practice, even if some forms will not occur as frequently as others. Consider for instance the following claim by Collins (1985, 19) that

> replication of others' findings and results is an activity that is rarely practised! Only in exceptional circumstances is there any reward to be gained from repeating another's work. . . . A confirmation, if it is to be worth anything in its own right, must be done in an elegant new way or in a manner that will noticeably advance the state of the art.

From the above analyses it is, I think, plausible to conclude that the categories 2, 3, 4, 6, 7, 10 and 11 of table I can be filled with historical experimental episodes. Categories 5 and 9 are most probably empty. After all, the claimed range of these types of reproducibility would presuppose an unrealistic continuity of belief in the correctness of the theoretical interpretation and result, as well as an unrealistic stability of the material and social conditions required for the material realization of the experiment. For the same reason, also category 1 is probably empty. (Perhaps a few crude material realizations—for example, producing static electricity by combing one's hair—might approximate the required range). Categories 8 and 12 are empty by definition, since the laypersons in question are presumed not to be theoretically informed.

As the above quotation demonstrates, Collins's discussion focuses on reproduction by other scientists, by means of new and better, and thus different, experimental procedures. In other words, he mainly deals with categories 9 and 10, the activities for which I, in conformity with his usage, have retained the term "replication." However, since also the categories 2, 3, 4, 6, 7, and 11 may be exemplified by various, concrete experimental episodes, we may conclude that the scope and role of experimental reproduction in actual practice is much larger than Collins assumes it to be.

2.4 NORMATIVITY, STABILITY, AND NONLOCALITY

So far I have argued for the claim that many experiments in the practice of the natural sciences are reproducible in one or more of its senses. At the same time a number of qualifications were added to this claim. First, I have pointed out the types of problems and failures that may and do turn up in actual attempts at reproducing experiments. Next, the essential fallibility of any particular attribution of reproducibility was noted. It also appeared that the question "reproducible by whom?" may be answered differently, depending on the relevant experiment and on the type of reproducibility. Finally, I emphasized the fact that reproducibility does not result from a passive registering of regularities but that it demands theoretical and practical work in order to get it.

Normativity

In addition to the factual occurrence of reproductions, we may ask whether or not reproducibility, in one or more of its forms, functions as a *norm* of experimental practice. I think it does (see also Hones, 1990). But, of course, this does not entail that every experiment should in fact be repeated by different scientists. In many cases its reproducibility is taken for granted. Yet, as we have seen, aspects of the norm of reproducibility have been incorporated into experimental practice through procedures such as the division of labor between laypersons and experimenters and the testing for (ir)relevant factors. Moreover, the operation of this norm becomes clear when the experiments are ambiguous, controversial, or very significant. Boyle, for example, thought the replication of his air-pump experiments by other researchers desirable, and he even expressed despair when many attempts turned out to be failures (see Shapin and Schaffer, 1985, 59–60). Faraday, in his experiments on magnetic rotation in the early 1820s, designed a more robust "pocket" rotation device and even sent it to some of his fellow scientists in order to facilitate the reproduction of his experiments by them (see Gooding, 1985, 120–122). And in the case of Weber's gravity-wave experiments the significance the involved scientists attached to replication was obvious (see Collins, 1975).

The argument so far is, I think, sufficient to refute the claim made by a number of authors that norms (or rules, standards, criteria, guidelines) that exceed the context of their local use do not play any role in scientific practice (see, e.g., Rouse, 1987, 93 and 119–125; Woolgar, 1988a, 45–48; and cf. Latour and Woolgar, 1979, 24). These authors reason as follows: applying a norm requires locally situated judgments; hence, norms do not and cannot determine practice; therefore, nonlocal rules or norms do not play any guiding role in science; instead, they merely form post hoc rationalizations of already established practice. However, although the first two statements are correct (or even trivial), the third does not at all follow. The claim that a rule should *compel* particular courses of action in order to have any practical impact at all is mistaken. The much more plausible alternative is, of course, that nonlocal rules may be practically effective *along with* all kinds of local factors.[25] As we have seen, the requirement of reproducibility was obviously consequential in such diverse contexts as that of Boyle, Faraday, and the gravity-wave experimentalists, even if it did not determine the particular courses of action.

Next, following upon the empirical issues of the occurrence of reproductions in scientific practice and of the actual functioning of the norm of reproducibility among scientists, there is the normative question of whether or not experiments *should* be reproducible. That is, can reproducibility be taken as a necessary condition of epistemic success, and do independent reasons for advocating such a norm exist? Again, the differentiation between the three types of reproducibility is required to obtain a sensible answer. It seems to me that the reproducibility of the material realization ought to be generally satisfied. In other words, I propose the norm of the reproducibility of material realizations as a constitutive norm of (good) experimental natural science. My main argument for this is that, without such a material stability, performing experiments and using them in technologies would be unreliable and dangerous.[26] With respect to the other two types of reproducibility, the situation is less clear. In general, reproducibility under a fixed theoretical description will facilitate communication among scientists, but it will also block creative reinterpretations of the experiment in question. Therefore, it does not seem advisable to put this type of reproducibility as a general constraint on experimenting. To a lesser extent the same applies to the case of replication. As we have seen, successful replication by means of a number of different experiments will establish the experimental result more firmly and will increase the systematic character of our scientific knowledge. Yet, in this case, too, completely different interpretations of this result can and should not be excluded. It would be inappropriate, therefore, to argue for replicability as a general norm of experimentation in the natural sciences. In sum, the norms of reproducibility under a fixed theoretical interpretation and of replicability of an experimental result are best regarded as regulative, rather than constitutive, norms.[27]

Stability

By reproducing an experiment or experimental result, it becomes more stable, that is, less sensitive to locally varying factors. As we have seen, reproducibility is a complex concept. The same holds good for stability: distinguishing types of reproducibility leads to different forms of stability.

A reproducible material realization requires that laypersons or other experimenters who possess or are able to acquire the necessary skills and who are willing or can be made to do the required craftwork can successfully carry out the experiment. Therefore, that an experiment is reproducible in this sense implies that these craft skills are learnable and, thus, that the experiment is more stable against, and less sensitive to, local variations in individual skillfulness. Moreover, in this case there is an important second form of stability, since local controversies concerning the theoretical interpretation of the experiment in question do not matter as long as they allow for reproducing its material realization.

Another form of stabilization works by means of replications. If an experiment can be replicated in a number of very different situations, the plausibility of its result will increase accordingly. The result will become less sensitive to criticisms of any particular experiment. That is to say, the more systematic the position of this result within a body of knowledge, the more stable it is. Apart from this, many replications aim at *increasing* the stability of an experimental result, in the sense that the effect is amplified, the noise reduced, or the result made less sensitive to potential disturbances. For instance, such a process of replicative standardization occurred in the case of Boyle's air pump: during the 1670s air pumps became routinely applicable and commercially available experimental devices (see Shapin and Schaffer, 1985, 274–276).

Finally, consider the reproduction of an experiment under a particular theoretical description. An example is a pneumatic experiment performed by Boyle first at home and later in the rooms of the Royal Society. Such a repetition stabilizes the experiment under the description in question in the sense that it turns out to be practicable at different times and places. Put differently, even when the experiment involves specific skills and a specific interpretation, it still shows a nontrivial insensitivity to variations in space and time.

We may conclude from the discussion in the present section that both the norms of reproducibility and the stable *aspects* of the experimental knowledge produced by their application can be seen to be specific, nonlocal patterns of scientific practice.[28] This is not to say that the knowledge obtained has been decontextualized or universalized but rather that it does, or has been made to, apply to a more or less large number of local situations that may vary in certain respects. In *these* respects the experimental knowledge has been delocalized.[29]

2.5 THE EXPERIMENTERS' REGRESS

After the analysis of scientific experimentation in sections 2.2 and 2.3, and after the immediate philosophical conclusions drawn from it in section

2.4, in the remaining sections of this chapter, I will apply the results obtained so far to three more specific issues. In the present section I will discuss and evaluate the so-called experimenters' regress, an argument originally put forward by Collins.

According to Collins, the experimenters' regress is at the heart of his (social constructivist) interpretation of science (Collins, 1985, 84). This regress concerns the replication and replicability of experimental results. Suppose that an experimental result q is, for some reason or other, controversial, and suppose that we try to resolve the controversy by replicating the experiment. Then, we may be confronted with the following circle or regress:

- whether or not q is believed to be true depends on the result of a correct q measurement. But:
- whether or not a q measurement is correct depends on whether or not q is believed to be true.
- and so on.

Collins concludes from this that in cases where there is no consensus about the correctness of a q measurement, replicating the experiment cannot bring about the consensus either. In these cases, the similarity of the second experiment and the competence of the second experimenter are particularly likely to be questioned. This means that replication cannot function as an objective, or purely empirical, test (of the truth) of a claimed experimental result. Yet, in practice the experimenters' regress *is* stopped at some stage. As a result of idiosyncratic, social processes of negotiation, experimenters come to agree on what counts as "similar experiments" and who as "competent experimenters." Thus, controversies concerning replication are not resolved by applying explicit, objective criteria, but rather by learning the relevant skills and being enculturated in a local community of experimenters.[30] In support of his claims, Collins offers, among other things, an extensive empirical study of the history of the replications of experiments concerning the existence or nonexistence of gravity waves (Collins, 1985, ch. 5).

For purposes of further discussion, it is helpful to reformulate the experimenters' regress on the basis of the analysis of experimentation presented in sections 2.2 and 2.3. Suppose, the original claim q results from the overall experimental process theoretically interpreted as $p \Rightarrow q$; and suppose that this claim is criticized on the basis of a replication: $p' \Rightarrow$ not-q. Then, two different ways are open to challenge this replication. The original experimenter may argue that the experimental arrangement described by p' is not an adequate q meter; that is, not-$(p' \Rightarrow q)$. Alternatively, he or she may claim that p' would have resulted in q had the experiment been competently performed. In fact, so this claim continues, it is not p' but, say, p'' that has been materially realized, where here indeed: $p'' \Rightarrow$ not-q.

These two ways to challenge the (dis)confirming power of replications lead to two different formulations of the experimenters' regress, which both occur in Collins's book, although they are not explicitly distinguished. The first can be found in the gravity radiation case:

> What the correct outcome is depends upon whether there are gravity waves hitting the Earth in detectable fluxes. To find this out we must build a good gravity wave detector and have a look. But we won't know if we have built a good detector until we have tried it and obtained the correct outcome! (Collins, 1985, 84)

The second formulation says that

> the *experimenters' regress* . . . is a paradox which arises for those who want to use replication as a test of the truth of scientific knowledge claims. The problem is that, since experimentation is a matter of skilful practice, it can never be clear whether a second experiment has been done sufficiently well to count as a check on the results of the first. Some further test is needed to test the quality of the experiment—and so forth. (Collins, 1985, 2)

For instance, in replications it may be that p and p' are not totally different but have one or more procedures (described by, say, p_1) in common. Therefore, when this is the case, the original experimenter may claim that in the second experiment p_1 has not been skillfully performed, because some tacitly assumed but crucially important aspects have been overlooked.[31]

The Experimenters' Regress Reconsidered

How to evaluate the experimenters' regress and its philosophical implications? I will submit six points of critical comment. It will become apparent that reconsidering the regress cannot mean refuting it. Nevertheless, it will also turn out that scientific practice does include a number of "stabilizing procedures" by which the effects of the regress are significantly *mitigated*.

The Knowers' Regress. Let me first note that an analogous "knowers' regress" is a characteristic of *all* methods of human knowledge acquisition. Since the only access we have to reality is through our methods of knowledge acquisition, claims about reality or truth and claims about the adequacy and/or the competent operation of our methods are, in one way or another, intrinsically dependent on each other. There is a certain holism here, which in another tradition has been described as a "hermeneutical circle" (cf. Hesse, 1986). For logical reasons, this general circle cannot be broken and as a consequence some kind of circularity is unavoidable.

This argument implies that, remarkably enough, the experimenters' regress bears no *specific* relation to experimentation. This becomes especially apparent from the first formulation of the regress, which says simply that a correct result presupposes a correctly followed method. The correctness of a particular method may be questioned independently of whether nonformalized skills are employed in applying it. After all, one may simply argue that the method in question is inappropriate to the study of certain domains of reality. In this sense some feminists, for instance, claim that the experimental approach to nature is inadequate because it is inherently masculine and oppressive.

However this may be, it is the second formulation of the regress that is the most basic in the view of Collins. Because experimental skills are "invisible" and cannot be "fully explicated, or absolutely established" (Collins, 1985, 129), in replications the problems of reaching an "objective" consensus about the correctness of the method employed are, according to him, aggravated to the point of becoming insoluble in principle.

Material Stability. Although some degree of hermeneutical circularity is unavoidable, the flexibility in experimental practice is reduced by the nontrivial requirement of the reproducibility of the material realization. As we have seen in section 2.3, it is true that reproducing experiments in this sense is possible even in the case of controversy about the theoretical interpretation. After all, one and the same reproducible material realization can be compatible with a class of possibly radically different interpretations. The point I want to make here is, however, that the reproducibility of the material realization is not affected by the existence of a controversy about its theoretical interpretation. Thus, this type of reproducibility entails a form of material stability of experiments by circumventing the experimenters' regress.

Collins, in contrast, focuses on replication as a test of theoretical hypotheses, such as the claimed existence of gravity waves. In doing so, he underexposes the significance of the processes of experimental action and production. This point is well illustrated in a section where he discusses a study of experiments to detect solar neutrinos. What is eventually produced in these experiments is "a wiggly line" on a sheet of paper, but Collins's claim is that the meaning of the experiments lies *exclusively* in the theoretical interpretations of such a curve:

> The point is this: there is little future in the scientist's reporting that $100,000 has been spend on sinking an instrumented tank of per-

chlorethylene in a gold mine to produce a wiggly line. No one is interested in a wiggly line. But, by virtue of its complete vacuousness, the claim that a wiggly line has been seen is unlikely to be contested. It will change no one's life; it will alter no networks of relationships. (Collins, 1985, 137)

In general I cannot agree with this. In fact, the fast and controlled material realization of wiggly lines of various forms on a monitor or on paper is what I am doing right now, by means of my word processor. Many people are extremely interested in such wiggly lines, and the large-scale introduction of these apparatuses has already changed a lot of lives. Of course this does not imply that the meaning of experimental science can be *reduced* to such material realizations and that theoretical interpretations and debate are of no significance. The point is, however, that the stable and reproducible material realization of experiments is a significant achievement of experimental practice. This achievement is, moreover, crucial to the potential utilization of experimental results within technological systems (see section 2.7).

The latter point is also relevant with respect to one of Collins's own case studies. For, in contradiction to the statements quoted above, in his laser study he readily accepts "the capacity to vaporize concrete" as a legitimate criterion to determine wether a laser works (see Collins, 1985, ch. 3). In Collins's line of thought, someone then might ask: who is interested in vaporizing concrete? Again, *my* answer would be: many people are and especially the military, whose role is not discussed by Collins but who played and continue to play such a decisive part in the whole development of laser theory and laser building (see De Ruiter, 1992).

Stopping the Regress. Despite the fundamental nature of the experimenters' regress, apparently, in practice it *is* stopped at some point by the acceptance of some view of what the result q is and what q measurements have been correctly performed:

> there was never any doubt that the laser could be replicated and never any doubt when it had been replicated. The fact remains that our experience of nearly all natural phenomena is like the experience of laser building; we know that the familiar objects of science are replicable. (Collins, 1985, 127)

According to Collins, the reasons for this acceptance are nonobjective and entirely social in character, since successful replications have a consensual rather than an empirical basis. In particular, he emphasizes the role played by experimental competence and tacit skill knowledge, which is acquired in a process of enculturation in an "experimental form of life."

I think that Collins is right in that nonformalized skills play an important and noneliminable part in science, especially experimental science. Consider again the operationalization of the notion of material realization through division of labor between a layperson and an experimenter. Also in this case, success in reproducing the material realization requires the possession or learning of certain manipulative skills on the part of the layperson. This learning process may be guided by explicit rules and knowledge, but it is certainly not determined by them. Yet, in spite of my agreement so far, I want to add a number of important qualifications to the role of tacit knowledge in experimental science.

Explicit Knowledge. In general Collins's view on the role of the inarticulable in science is too one-sided to be true. He states as a central conclusion that:

> Proper working of the apparatus, parts of the apparatus and the experimenter are defined by their ability to take part in producing the proper experimental outcome. Other indicators cannot be found. (Collins, 1985, 129, italics omitted)

Consequently, explicit, prospective design of the experimental setup does not play a role in the definition of its eventual success. The proper working of an experiment can only be established retrospectively. For instance, a TEA-laser (or transversely excited atmospheric gas laser) works because it is able to produce the proper outcome of vaporizing concrete.

Yet, the success of a TEA-laser in meeting this criterion is not the only "explanation" of its working. A satisfactory explanation will refer to the skillful experimental action and production as well as to the prospective design characteristics of the device. Such explicit, theoretical indicators of success *can* be found, and that even in Collins's own account of the case. For example, it was known in advance that building a high-power laser requires a relatively high gas pressure, a high voltage, and an accurate construction of the anode and cathode; also, the gas should have a definite and well-known composition, and it may not contain any air or other contamination (Collins, 1985, 51–54). Characteristics such as these cannot be excluded from the definition of a working TEA-laser. Without this explicit knowledge, no amount of skill would have been able to produce a working device.

This objection can be related to the two formulations of the experimenters' regress discussed above. As I observed, only the second formulation refers directly to skills. The first formulation concerns the question of what is considered to be a good q meter. As I argue here, the answer to this question is, in part, dependent on explicit, theoretical knowledge.

Delocalization. Next, remarkably enough, it is the very procedure of replication of a result q by means of different experimental processes that entails a certain form of "delocalization." If a replication is successful, q becomes less dependent on any one of the specific experimental processes p, p', p'', etc., and therefore also less dependent on the specific skills that are necessary to produce these particular processes. In replicating experiments, the scientists *abstract* from (a part of) the local situations in which q has been produced. Therefore, the meaning of a replicable experimental result, in some way, transcends the separate contexts in which it has been produced.[32] The successfully replicated result q attains a more systematic position, mostly within the context of a body of theoretical knowledge. This is clearly illustrated in the replications of Avogadro's hypothesis, mentioned in section 2.3.

Collins does not deny the occurrence of this kind of delocalization, but he does not fully account for its meaning and role in science either. As mentioned above, it is the experimenters' regress upon which he bases his interpretation of science. His discussion of the network character of scientific concepts (see Collins, 1985, ch. 6) is valuable, but he does not feed it back into his earlier conclusions concerning the more local experimental episodes (cf. also Hesse, 1986). If we take full account of the nonlocal aspects of experimental and theoretical science, some of Collins's views have to be revised. This applies especially to the claim that the proper working of a particular experiment and the proper expertise of a particular experimenter can be exclusively derived from their ability to produce the correct outcome of this one particular experiment.[33]

Standardization. A final point that is relevant to the issue of the experimenters' regress and the role of skills is that experimental practice itself shows a clear tendency towards standardization. An important stage in experimenting is making experimental processes and results more robust and therefore less dependent on the skills of specific individuals. Thus, standardization can be considered a specific kind of replication. Faraday's experiments on magnetic rotation in the early 1820s offer a striking illustration of the significance of standardization. Having successfully performed a first experiment for himself, Faraday immediately went on to build a more standardized, "pocket" version of his rotation device. Gooding (1985, 121) discusses this experimental episode and concludes:

> He made it easy for others to reproduce the effect by sending them the device. This also shows how aware he was of the possibility of failure. It reduced the risk by making it unnecessary for others to acquire all of the practical skills and tacit knowledge that Faraday had so laboriously built up.

Just like other kinds of replication, standardization is an important procedure by means of which experimenters attempt to delocalize their results. It is true that experimenting essentially involves particular, skillful action and production. But it is equally true that these activities do not exhaust experimental practice, a substantial aspect of which is trying to make the results reproducible by others. The continual reconstruction of knowledge for the purpose of making it more stable is an essential feature of experimental (and indeed of all scientific) practice (cf. Nickles, 1988).

In conclusion, the argument of this section has been that the reality and even the unavoidablity of the experimenters' regress does not imply that experimental results are entirely constituted by tacit knowledge or that their validity is necessarily restricted to their local production context. The reason for this is the existence of a number of stabilizing procedures, to wit: stabilization through reproducing the material realization; stabilization through explicit, theoretical design; stabilization through replication; and stabilization through standardization.

2.6 DATA VERSUS PHENOMENA?

For the purpose of further clarifying the preceding account of experimentation and reproducibility, it is instructive to briefly examine its implications for the distinction between experimental data and phenomena. The importance of this distinction has recently been emphasized in an illuminating paper by Bogen and Woodward. Their main philosophical conclusion concerns the relation between explanation and observation. They claim that systematic scientific theories do not purport to explain the observable data but only the nonobservable phenomena:

> Data, which play the role of evidence for the existence of phenomena, for the most part can be straightforwardly observed. However, data typically cannot be predicted or systematically explained by theory. By contrast, well-developed scientific theories do predict and explain facts about phenomena. Phenomena are detected through the use of data, but in most cases are not observable in any interesting sense of that term. (Bogen and Woodward, 1988, 305-306)

Bubble-chamber photographs and thermometer readings are examples of data. The existence of phenomena, such as weak neutral currents or the melting point of lead, is inferred from the data in combination with other (e.g., statistical) premises.

I will not discuss Bogen and Woodward's main philosophical claim but instead focus on the rather sharp contrast they make between data and phenomena (Bogen and Woodward, 1988, 305-322). Data, they claim, are *observable*; they are *created* by experimenters; and they are highly dependent on the *multicausal* and *local* experimental contexts in which they have been produced. Phenomena, in contrast, are inferred and are not observable; in general, they are real and not created; their occurrence results from the interplay of a relatively small number of causal factors; and they can be detected in a variety of experimental contexts.

Now, how does this distinction between (experimental) data and phenomena relate to the analysis presented in this chapter? Suppose we have a theoretical description of the form $p \Rightarrow q$. Then it seems to me that the experimental result q will, in many cases, provide a description of what Bogen and Woodward denote as phenomena, while what they call data will be included in the composite theoretical premise p. After all, the experimental result, just as the claimed existence of a phenomenon, is the outcome of a process of inference from the theoretical description p. However, if this interpretation of Bogen and Woodward's notions of data and phenomena is valid, the above analysis of experimentation implies that their sharp contrast between data and phenomena cannot be upheld.

In the *first* place, it is not the case that the data included in the theoretical description p of an experiment are all or even for the most part observable in Bogen and Woodward's sense, in which what is observed "should figure somehow in retinal interactions" (Bogen and Woodward, 1988, 347). For instance, even when we want to detect such a simple phenomenon as the boiling point of a liquid, we have to make use of, among other things, the data that "the liquid is pure" and "the air pressure remains constant during the measurements" (see Radder, 1988, 60–69). In this case these crucial data are by no means straightforwardly observable in the required sense. Since it is not difficult to find many more counterexamples, especially in more advanced experiments, the claim of Bogen and Woodward is simply not plausible as it stands.[34] From my point of view, it would rather be the material realization of the experiment that is "observable." However, its description in common language can definitely not be used to infer the experimental result. For this purpose we do need a *theoretical* description including theoretically interpreted data.

Second, if data are "created" in the laboratory,[35] a fortiori the same must apply to phenomena. After all, the specific realization of a phenomenon (e.g., the boiling point of a liquid) depends crucially on the realization of the created situations described by the data (e.g., that the liquid is pure). Bogen and Woodward's claim to the contrary seems to rest on a confusion of epistemological and ontological matters. They are right in maintaining that many phenomena are replicable; that is, they can be shown to occur in a variety of experimental situations. In these cases we have good epistemological reasons to conceptualize the phenomenon or experimental result "as such" and ask for an explanation of it in abstraction from the various ways in which it has been experimentally produced. For instance, if we are successful in experimentally replicating a fixed boiling point for our liquid, we may plausibly proceed and look for a systematic explanation of it in terms of a molecular theory of phase transitions. Hacking's view of the creation of phenomena, however, is ontological. He claims that a phenomenon, such as the Hall effect, does "not exist outside certain kinds of apparatus" (Hacking, 1983b, 226). This claim cannot be refuted by pointing out that when a phenomenon is replicable, the phrase "certain kinds" may be broadened into "a variety of kinds," since any particular realization of the phenomenon remains dependent upon *some* specific set of laboratory conditions.

Third, Bogen and Woodward claim that not only phenomena (1988, 317) but also data should be reproducible.

> Data must also be such that it is relatively easy to identify, classify, measure, aggregate and analyze in ways that are reliable and reproducible by others. (Bogen and Woodward, 1988, 320)

But, how can these requirements be met if data are at the same time "idiosyncratic to particular experimental contexts, and typically cannot occur outside of those contexts" (Bogen and Woodward, 1988, 317)? Thus, there is a tension in Bogen and Woodward's view between the reproducibility of the data and the reproducibility of the phenomena. This tension is due to the claimed locality and idiosyncracy of data as contrasted to the nonlocality and stability of phenomena. In my view, both data and phenomena may be reproducible and may therefore be nonlocal and stable in certain respects. Yet, as we have seen in the preceding sections, the form this nonlocality and stability takes may be different in the two cases as a consequence of differences in type of reproducibility.

From the point of view of this chapter we can, I think, conclude that Bogen and Woodward have the right intuition that there is a substantial difference between reproducibility under a fixed theoretical description $p \Rightarrow q$ (where p includes the data) and the replicability of the phenomenon or experimental result described by q. The questions "reproducibility of what?" and "by whom?" may get different answers in the two cases. As a consequence, these types of reproducibility play varying roles in scientific research. Nevertheless, it will also be clear that these differences do not entail the sharp contrast between data and phenomena that has been claimed to exist by Bogen and Woodward.

2.7 EXPERIMENTAL SCIENCE AND ITS SOCIAL LEGITIMATION

A final question to consider is this: How do experiments that have been tried out successfully by individual experimenters acquire a more widespread public recognition? This question has been investigated by some historians of experimentation recently. Gooding, in his studies of the work of Faraday and his contemporaries, points to the significance of public demonstrations of experiments with the aim of making particular phenomena and processes seem evident and natural. By being *witnessed* by lay audiences, experimental results would become recognized as part of a collective body of knowledge (Gooding, 1985; 1989, 191 and 202–203). At a more fundamental level, there is the question, not so much of the public recognition of *specific* experimental results, but of the experimental method as such. It is this question of how experimental natural science is socially legitimated that is at issue here.

For the days of the early Royal Society, this issue has been discussed in detail by Shapin and Schaffer (1985, 55–60; cf. also Shapin, 1988). They emphasize the central role of witnesses in generally establishing the authority of the experimental way of producing knowledge. These witnesses testify to the fact that the experiments in question have really been performed and that the knowledge produced is reliable. For this purpose witnesses should be credible, which was taken to mean that they should originate from the same (higher) classes of society as the natural philosophers themselves did. In this way, the new experimental philosophy opposed all at once the secret experiments of the alchemists, the thought experiments of the rationalist scientists, and the antiexperimentalism of the school philosophers.

Yet, although the early British experimentalists were successful in establishing the experimental "form of life" as a social phenomenon, it will be clear that this procedure for legitimizing experimental science has a rather restricted scope. In fact, even in the beginnings of the Royal Society, the procedure involved two ambiguities. These concerned who should witness and what should be witnessed. Should witnesses be laypersons who have no real connection to experimental philosophy, or should they "have knowledge of the things they deliver," as Boyle once put it (Shapin and Schaffer, 1985, 59)? And should they testify only to the existence of "matters of fact" or also to the truth of experimental hypotheses? In practice, all different possibilities occurred. It will be clear, though, that this way of legitimizing experimental science leads to a dilemma. On the one hand, when witnesses are knowledgeable and are required to validate experimental hypotheses, they are likely to be prejudiced in favor of one claim or the other. Consequently, their role as independent and credible testifiers may be questioned. As Dear (1985, 156) has argued, this problem was to some extent recognized by the members of the Royal Society. On the other hand, when witnesses are genuine laypersons required to establish matters of fact, their testimony cannot be of any help in settling controversies about different theoretical interpretations of the same matter of fact, such as the debate between Boyle and Hobbes on the void-in-the-void experiment. What is more, in the overwhelming majority of modern experiments, the theoretical interpretations will simply be incomprehensible to such laypersons.

In conclusion, the procedures for validating experimental knowledge employed by the Royal Society involve an inherent and irresolvable tension. Moreover, it will be clear that modern experimental science is no longer socially legitimated by means of the method of direct witnessing. The modern laboratory is much more a private place than its seventeenth-century precursor (see Shapin, 1988, 404). Nowadays, debates on the interpretation of scientific experiments take place exclusively among the scientists themselves. This is, of course, not to say that researchers may not make use of all kinds of resources from the society at large. But it means that, in general, people in society do not themselves actively *participate* in the technical-scientific debates.[36]

Experimental Science and Technology

Given this state of affairs, the question arises as to what social legitimation has replaced the early procedures of the Royal Society. Why is society willing to provide ever larger means for an experimental science that has become ever less intelligible to the general public? To be sure, one cannot expect this question to have only one answer. Different times, different places, and different sciences may have produced different legitimations.

Yet, concerning experimental natural science as such, I think two legitimations play a fairly prominent and general role. The first is the claim that science is valuable because it delivers the truth about nature or, at least, promises to eventually give a true account of nature. The second major social legitimation is framed in the claim that experimental science is practically useful, that its results can often be fruitfully incorporated into all kinds of technological projects. Gooding, for instance, mentions the fact that, already in the 1820s, British experimenters on electromagnetism were eager to stress the significance of their experiments for navigation and the discovery of ores (Gooding, 1989, 202–203). Actually, in present-day society, the "technological" legitimation seems to be the most influential, and I will restrict myself to it in the following. As a matter of terminology, I will speak of *experimental technology* in cases where materially realized scientific experiments are used as parts of technological systems.[37]

So far this answer to our question seems rather evident, but it is also rather uninformative. The more difficult task is to specify the notion of social legitimation through the technological potential of science. My suggestion is that a considerable part of the social legitimation of experimental natural science rests on the possibility of actually achieving reproducible material realizations of experiments within the context of experimental technologies. That is, it is due to the action and production aspects of experimenting and to the availability of *some* theoretical description on the basis of which we can obtain reproductions. What is not required is a general agreement on specific theoretical interpretations. Consider, for instance, the manned U.S. space missions to the moon in the 1960s. In this case, as long as the experimental technologies continued to work in the sense of remaining materially realizable in a stable and reproducible manner, the average politician and taxpayer did not at all bother about possible controversies concerning the correct theoretical interpretation of some feature of the project. For example, a debate on the question of whether the trajectories of the spaceships should be calculated by means of classical mechanics or by means of the special theory of relativity does not seem to be relevant to the social legitimation of this experimental technology.[38] As we will see now, this view has some significant consequences with respect to the role of experimental science and technology in present-day society.

Devices and Black Boxes

On the basis of the previous analyses, we can also provide a general explanation of the *possibility* of the worldwide use of experimental science in technological projects. Given the complexity and the spatiotemporal variability of theoretical interpretations, the relative autonomy of the material realization with respect to particular theoretical descriptions is a crucial and fortunate feature both for producers and for consumers of experimental technology. On the one hand, if it were the case that first a general, scientific consensus must arise about all the details of a particular account of how an experiment works, then in view of the current state of natural science, its technological adoption would be difficult. On the other hand, the large-scale use of experimental technology would become virtually impossible if every user should first acquire and accept the knowledge about even one of its theoretical interpretations. Thus, it is this relative autonomy of reproducible material realizations that makes it possible to produce and use experimental technologies on a large scale, even in the face of theoretical controversy and theoretical ignorance.

This explanation of the possibility of experimental technology can be fruitfully used in commenting upon some other recent views on the issues under discussion, notably those of Borgmann and Latour. Borgmann, in his book *Technology and the Character of Contemporary Life*, on the one hand, fully accepts what he calls the "apodeictic" results of natural science, and he interprets these results in a strongly realist way (see Borgmann, 1984, 15–31). On the other hand, he is very critical of the form modern, science-based technology has taken in our present-day society. He characterizes this form as a "device paradigm" (Borgmann, 1984, esp. ch. 9). A device, such as a central heating system, is a technological artifact that consists of machinery and that procures a commodity. One of the central characteristics of modern technological devices is that their machinery is hidden from, and their operation unfamiliar to, their users. This admits of a certain variability of the machinery as long as the device continues to provide the same commodity. The core of Borgmann's criticism of modern technology, then, is that this paradigmatic device character disengages and alienates people from their ontic roots. According to him, it makes them incapable of a truly human interaction both with their natural and with their social environment.

If we leave aside Borgmann's inadequate views on science, we see that the analysis of experimental science along the above lines goes some way in explaining the device character of a technology that makes use of an immense body of specialized, fragmented, and ever changing experimental knowledge and artifacts. As we have noted, demanding an intensive engagement of every user with this body of knowledge would make experimental technology practically impossible. For large-scale experimental science and technology, a division of theoretical and material labor is a crucial requisite. As a consequence, for the average consumer, modern experimental technologies will necessarily have the form of devices or, to use a related term, black boxes.

Next, this relation between the notions of "device" and "black box" enables me to make some comments on Latour's views on the development of, to use his term, technoscience (see Latour, 1987b). He interprets this development as essentially a struggle for power within networks of heterogeneous, antagonistic "actants." In this process various powerful actors, or Machiavellian "Princes," strive for control and stability of their networks. The more black boxes a Prince is able to uphold, the more stable his power. According to Latour, it is a defining characteristic of a technoscientific black box that well-established scientific facts and unproblematic technological artifacts are intrinsically connected in it (Latour, 1987b, 131). Put in my terms, Latour's view implies that a stable, experimental technology requires both a general, be it an implicit, agreement on its theoretical description and a reproducibility of its material realization.

On the basis of the present analysis, such a view appears not to be fully adequate. It is true that theoretical controversies that question the reproducibility of the material realization—and thus the working—of an experimental technology will affect its stability. In contrast, all theoretical interpretations that are compatible with the same, reproducible material realization are in this respect[39] equivalent, so that a controversy about these interpretations will not directly touch the stability of the technoscientific black box. In other words, in many cases, experimental technologies are black boxes in the original sense of the word: they are systems for which only input and output matter, *independent* of whether the account of their internal workings is scientifically well established or controversial.[40] The important implication of this is that it enables us to analyze and evaluate the worldwide pervasion of materially realized experimental technologies without having to assume the existence of Latourian super-Princes, who make all the elements of the technological black boxes "act as one" (Latour, 1987b, 131). The imperialistic expansion of technological devices is compatible with a considerable amount of pluralism at the level of their theoretical interpretations.

Revaluing the Invisible Actors

In concluding this chapter, I want to return one more time to the early days of the Royal Society. My view so far has been that the current social legitimation of modern experimental science cannot be based on the witnessing of credible lay participants. Instead, it rests (at least partially) on the potentialities of materially realizing experiments in a stable and reproducible way. One of the goals of this chapter has been to make explicit these crucial action and production aspects of experimenting by means of the notion of material realization and to point out its actuality in processes of division of labor.

In the context of the Royal Society, this means revaluing the importance for legitimating experimental science of the "invisible actors," such as Boyle's many laborers, operators, assistants, and chemical servants (see Shapin, 1988, 395–396; 1989). This implies a shift to an even stronger form of participation of laypersons: from participation through witnessing to participation through performing the experiments and producing the experimental systems. In this way, the ambivalence of the Royal Society procedures for legitimizing knowledge, noted at the beginning of this section, has been resolved. In the struggle to establish particular theoretical interpretations, only knowledgeable scientists participate. The legitimation of experimental knowledge claims, however, does not presuppose the truth of a specific interpretation. Instead, it derives from the possiblity of the reproducible material realization of the experiments by "lowly folk," whose performance is stable just because it "stands above" any of the particular, competing interpretations (cf. Dear, 1985, 156; see also Radder, 1988, 74–75).

More in general, the argument of this chapter entails *both* a critique of a theory-biased philosophy of science (which has its roots in a much broader, cultural bias toward intellectual work) *and* a revaluation of the mostly invisible but crucially important craftwork in the experimental natural sciences, as it is performed by the experimenters themselves or by what I have called laypersons.

CHAPTER 3

Heuristics, Correspondence, and Nonlocality in Theoretical Science

3.1 INTRODUCTION: INTERTHEORETICAL CORRESPONDENCE AS A NONLOCAL PATTERN

Let us now turn from experiment to theory. What about the realization of theoretical nonlocalities in science? In this chapter, I will discuss this question by means of a study of the heuristic significance of the so-called generalized correspondence principle. More in particular, I will examine of what the generalized correspondence principle is a generalization, and I will make use of the results of this examination in historically and philosophically assessing the heuristic value of the principle. The following introductory observations may serve to define the subject.

The discussion will be based on the plausible assumption that developing new scientific knowledge is neither a completely algorithmic nor a purely arbitrary process. In other words: On the one hand, it is certainly not the case that there is (or might be invented) some algorithmic method, a set of universally valid and unambiguously applicable rules that would uniquely determine the generation of new knowledge. On the other hand, this fact does not exclude that in the practice of science certain heuristic rules are used, which guide the search for new knowledge by drastically restricting the number of possible roads or by positively suggesting which general directions to take in the searching process.

This assumption can be explained somewhat further by regarding the development of the sciences as a production process (cf. Bhaskar, 1978, ch. 3). In this view, expressions such as the "invention" or "discovery" of new knowledge are strictly speaking misleading. Gold can be discovered; knowledge cannot. Knowledge is realized in a production process. In this process, "raw materials" are transformed into a final product with the help of particular tools and under guidance of specific rules and goals, just as is the case in any other production process. Hence the "discovery" of new knowledge is better described as the successful transformation of existent knowledge. This is a process in which raw materials (e.g., older theories) are processed with the help of tools (e.g., experimental or computational devices) and that is governed by methodical and methodological rules (e.g., rules guiding the selection of random samples or implying a preference for consistency or simplicity) and by epistemic or social aims (e.g., empirical adequacy or the production of socially useful substances).

Probably no one will contest the fact that in such transformation processes, guiding factors are operative in one way or another. Of course, the crucial question is what exactly the nature and scope of these factors are. Many philosophers will consider them relevant only if they exhibit at least some degree of nonlocality; in other words, if they are not exclusively applicable to that transformation process in which we have found them in the first place. Philosophical study of scientific heuristics, then, is the reconstruction and interpretation of nonlocal patterns of tools, rules, and goals guiding the production of new scientific knowledge.

Up to this point the characterization of the development of knowledge as a transformation process has been rather formal. If we want to get any further we will have to characterize scientific knowledge in a more substantial way. Regarding the development of the modern natural sciences, we should at least distinguish between *experimental*, *mathematical*, and *conceptual* aspects (in the following we will occasionally combine the latter two into the *theoretical*).[1]

As to the heuristics of experimentation, including observation, we may for instance examine how stable and reproducible phenomena can be realized (Hacking, 1983b, ch. 13; Radder, 1988, ch. 3; chapters 2 and 6 of this book), how experiments end (Galison, 1987), how consensus about new observations is reached (Gooding, 1986), or how innovations of scientific instrumentation arise (Price, 1984). However, the possible heuristic rules that guide these processes will not be dealt with here, since in this chapter I will restrict myself mainly to theoretical, that is, both conceptual and mathematical, innovations.

The heuristic strategies for theory development that are recommended in the literature roughly refer to two levels. The heuristics of articulating or elaborating a theory, paradigm, or research program is often discussed in terms of the role of models and analogies (Hesse, 1963), of metaphors (Boyd, 1979), and of the de-idealization of models and hypotheses (Lakatos, 1970; Shapere, 1977). Furthermore, various studies exist that stress the heuristic significance of intertheoretic relations in the replacement of entire theories, paradigms, or research programs. Here attention is paid frequently to the heuristic value of forms of reduction (Nickles, 1973) and to requirements of symmetry, of invariance, and of the conservation of conservation laws (Post, 1971; Redhead, 1975); finally, the heuristic importance of the generalized correspondence principle is emphasized (Post, 1971; Krajewski, 1977; Zahar, 1983; Fadner, 1985). In the following I will focus on the latter.

The term "correspondence principle" springs from atomic physics. In 1920 it was introduced by Bohr in the context of a procedure to tackle certain problems in the field of atomic spectra. In view of the fact that the account of the spectra given by the classical physical theories (mechanics and electrodynamics) had proven to be empirically wrong, Bohr tried to construct a new, quantum theoretical explanation. In doing so he was led by the idea that for certain domains of phenomena, classical theory and quantum theory should lead to the same results; or, in other words, that in these domains there had to be a "correspondence" between the theories in question. This requirement of correspondence functioned as an explicit, and moreover a rather successful, heuristic rule in constructing modern quantum theory (see, e.g., Meyer-Abich, 1965; Jammer, 1966).

However, although the *term* "correspondence principle" was introduced specifically in atomic physics and only in 1920, there certainly exist wider and earlier applications of the principle. This fact has led a number of philosophers of science to postulate the existence of a so-called generalized correspondence principle. It is claimed that this is applied regularly as a successful heuristic strategy, at least in the more "mature" sciences and especially in physics. Fadner (1985, 830), for instance, mentions a large number of such applications, starting with Newton and via Bernoulli, Young, Fresnel, Clausius, Maxwell, Planck, Einstein, Bohr, De Broglie, Heisenberg, and Dirac, ending with Watson and Crick.

In this chapter, I will examine in detail *what* is generalized if we generalize the correspondence principle from atomic physics and, consequently, what the heuristic power of the generalized principle does and does not amount to. In view of this, I will first, in section 3.2, sum up four accounts of the principle. In order to be able to discuss these accounts in sufficient detail I will furnish a precise reconstruction of the role of the principle in quantum theory, proceeding in two steps. In section 3.3, I will examine the various parts played by the correspondence principle during the genesis of modern quantum mechanics, around the years 1913 through 1925. Next, section 3.4 provides an analysis of the more normal-scientific ways in which the principle is used in all kinds of applications of modern quantum theory.

The results of this reconstruction enable me to further analyze and evaluate the accounts of the generalized correspondence principle described in section 3.2. Again, I will proceed in two steps. Section 3.5 mainly deals with an assessment of the different formulations of the principle. Its conclusion is that the principle should at least be considerably differentiated and qualified, especially regarding its nature and its range of applicability. This implies that the heuristic value of the principle also has to be reassessed. This will be done in section 3.6.

48   IN AND ABOUT THE WORLD

Finally, the assumption of a generalized correspondence principle also raises more general philosophical questions, namely through the problem of how its application may be philosophically justified. Here we meet with methodological, epistemological, and ontological presuppositions lying behind the principle. As we will see in section 3.7, various philosophical views on the matter are proposed. In the conclusion of this chapter, these views, as well as the related problems of theory reduction and the discovery-justification distinction, are briefly discussed.

3.2 THE GENERALIZED CORRESPONDENCE PRINCIPLE

In this section, I will summarize four accounts of the generalized correspondence principle: they will be, successively, the account by Post, by Krajewski, by Zahar, and by Fadner.

Post (1971) argues for a philosophical rehabilitation of heuristics. Science is not a blind process of trial and error. Not only the selection but also the proposed variations of scientific knowledge are guided by rules. In the first case, these rules are methodological, and in the second, they are heuristic. In his paper, Post confines himself to those heuristic rules that place *theoretical* constraints in advance on the construction of new theories as successors of older theories. Such rules are "conservative" in the sense that the accomplishment of the older theories is retained in their successors. It is in particular the generalized correspondence principle—in Post's view the most important heuristic constraint—that guarantees this. At first Post describes the principle as follows:

> any acceptable new theory L should account for the success of its predecessor S by "degenerating" into that theory under those conditions under which S has been well confirmed by tests. (Post, 1971, 228)

A typical example of such a correspondence is the relation between Newtonian classical mechanics and the theory of special relativity. In the limit of low velocities, the latter theory "degenerates" into the former. The new theory L shows why the old theory S did work in certain domains and not in others. Thus, L supplies an *explanation* of the success of S^*, the well-confirmed part of S. According to Post such progress from S to L never involves any losses:

> I shall even claim that, as a matter of empirical historical fact, L theories always explained the *whole* of S^*; that is, contrary to Kuhn, there is never any loss of successful explanatory power. (Post, 1971, 229)

Moreover, he emphasizes that the correspondence not only concerns the empirical predictions in the domains in which S was successful, but also applies to what he calls the lower theoretical structure of S. Even if the new theory offers a theoretical interpretation of the terms of S that is incompatible with the interpretation by S—which frequently occurs according to Post—the "structure" or "pattern" of the lower theoretical levels of S will be preserved. He adduces the periodic table in chemistry to illustrate his point: whereas the interpretation of a chemical element as an immutable entity has been abandoned, the ordering (the pattern) of the elements in the periodic system has remained the same.

The generalized correspondence principle operates as a constraint on L. It does not uniquely determine the transition from S to L. Post shows, with examples taken mainly from twentieth-century physics, that the principle has been successful in many cases. Remarkably enough, however, it is precisely the relation between classical mechanics and nonrelativistic quantum mechanics that does not satisfy the principle, according to him. Owing to the considerable normative weight that Post ascribes to the generalized correspondence principle, he thinks that this fact speaks rather against modern quantum mechanics than against the principle itself (Post, 1971, 233–234).

In his book *Correspondence Principle and Growth of Science*, Krajewski sees the generalized correspondence principle as a methodological rule that is characteristic of the development of mature sciences, especially of modern physics. He differentiates between a descriptive and a normative version of the principle. Although he focuses on a logical analysis of the descriptive version, he considers a normative use of the principle—both in a heuristic sense in advance and in a justificational sense afterwards—completely legitimate (Krajewski, 1977, 9 and 41).

His views on the generalized correspondence principle are explained by Krajewski with the help of Nowak's conception of science in terms of idealization and factualization (or concretization). In this context he defines the correspondence principle as a relation between an earlier, more idealized law or theory and a later, less idealized one (Krajewski, 1977, ch. 4). Such a correspondence relation obtains for instance between Boyle's gas law and that of Van der Waals. Given the idealizing assumptions that molecules are pointlike and that no intermolecular forces are present, the latter law transforms into the former. Also, under the conditions $h=0$ and $v/c=0$, respectively, nonrelativistic quantum mechanics and the theory of special relativity change into classical mechanics. More precisely, in the case of theories, it holds that under idealized conditions (to be noted as $p_i=0$) the basic equations of the later theory L can be carried over into those of the earlier theory S, mostly by way of some suitable limit procedure.

Krajewski attaches great importance to a *deductive* foundation of the correspondence relation. This relation, however, cannot be directly formulated in a deductive way, because as a matter of fact, L and S are often inconsistent. These inconsistencies are related to the fact that L assumes that $p_i \neq 0$ while S is based, implicitly or explicitly, on the premise that $p_i = 0$. Nevertheless, Krajewski claims that there does exist a logically sound deductive relation between slightly modified versions of L and S. In this indirect manner, he intends to supply as yet a deductive justification of the correspondence relation.

Two kinds of correspondence relations are distinguished by Krajewski: homogeneous and heterogeneous relations (Krajewski, 1977, ch. 5). In the case of homogeneous correspondence, the meanings of the terms from S and L are identical, in contrast to what is the case for heterogeneous correspondence. In the latter case, we need a "translation" of the vocabulary V_S of S into the vocabulary V_L of L. According to Krajewski, the correspondence between classical mechanics and special relativity is homogeneous. In his view the term "mass" has the same meaning in both theories, namely that of being the "measure of inertia." The difference between the two theories is merely that they propose differing synthetic statements concerning that same mass. The correspondence relation, then, claims that a common domain exists for which the basic relativistic equations transform into those of classical mechanics. An example of heterogeneous correspondence can be found in the relation between quantum and classical mechanics. Quantum mechanical observables are defined by means of operators and thus have, according to Krajewski, a meaning different from their classical counterparts. Nevertheless, a definite connection obtains between quantum and classical terms, and hence we meet with a (heterogeneous) correspondence relation also in this case.

When a correspondence relation exists between successive theories, the later theory L "includes" the older theory S in the sense explained above. L makes explicit the idealizing features of S, thus explaining the approximate validity of S in certain empirical domains. In this way, S is being reinterpreted in the light of L. The generalized correspondence principle can also be used heuristically in search for new theories that are increasingly less idealized and consequently ever more adequate from the empirical point of view (cf. also Kirschenmann, 1990). For that reason, Krajewski interprets the correspondence relation not only as a logical but also as a temporal relation.

Zahar sets out to answer the following two questions: "is there in physical science such a thing as rational heuristics, and, if so, how does it operate?" (Zahar, 1983, 243). The first question is answered by him in the affirmative, and his answer to the second is worked out within the framework of Lakatos's views on the development of research programs. According to Zahar, the hard core of a research program will contain metaphysical principles that, after being reformulated in prescriptive terms, simultaneously play a part in the positive heuristic as guidelines for further development of the program. The principle of the uniformity of nature, for instance, leads to the heuristic rule to strive for the unification of theories. Within scientific research programs, such principles and rules, which in Zahar's view originate from daily life, are made more concrete and precise (Zahar, 1983, 249–252). Another important heuristic rule is the correspondence principle, which Zahar defines as the requirement that a new theory should be able to explain the empirical success of its predecessor and, more in particular, that the laws of the newer theory L should under certain conditions (e.g., $h \to 0$ in quantum theory) change into their counterparts of the older theory S (Zahar, 1983, 246).

How then does such a rational heuristics operate? According to Zahar, in a deductive way and at a metalevel. The latter takes into account the fact that the theories S and L are often incompatible, so that there cannot exist a straightforward relation of deducibility between them.[2] In following rational heuristic rules, we derive certain true consequences from S (true, that is, in the light of L) and use them in a deduction of the new theory L. These deductions employ both the appropriate empirical data and the previously mentioned theoretical heuristic rules. Zahar's main example is the discovery of Newton's theory of gravitation, which required two steps:

> In the first stage Newton derived his laws of motion from the Cartesian law of inertia together with the Principle of the Proportionality of cause to effect; he also *deduced* an inverse square law for central accelerations from Kepler's laws. The second stage involves a meta-argument which resorts to the Correspondence Principle and to the Principle of Sufficient Reason in order to determine a new gravitational theory. (Zahar, 1983, 258–259)

The correspondence principle applies here in the sense that in the case of a solar system containing only one planet of a mass very small compared to the sun's mass, Newton's theory of gravitation would imply Kepler's laws. In other words, under these idealized conditions, Kepler's laws are true, even in the light of Newton's theory. In this way, Zahar claims to have solved the problem of

> how Newton could have derived his theory from laws contradicting it: Newton *reflected* on Kepler's laws without supposing them to be exactly true; he extracted certain logical consequences from them and went through a chain of deductive reasoning aimed at determining a proposition X such that X conforms to Newton's third law and also tends to Kepler's laws in certain limits. (Zahar, 1983, 259)

We see that Zahar advocates a remarkably strong form of heuristics. First, heuristic arguments are being viewed as strictly deductive; and second, they move from the older to the newer theory. In contrast to, for instance, Post who claims that heuristic rules merely place constraints on new theories, Zahar interprets these rules as enabling us (if supplemented by the relevant empirical data) to deductively infer the new theory from the old. Nevertheless, he maintains that a distinction between context of justification and context of discovery remains necessary, thus agreeing with Popper and Lakatos, among others. In their view, the data or principles that lead to the construction of a theory in the first place cannot at the same time play a part in its later justification.

As I mentioned in the introductory section, in *Fadner's* (1985) view, the correspondence principle is a general principle of natural science. He, too, distinguishes two ways of applying the principle, namely "deductively" and "inductively." In the first case, it is demonstrated after the conception of a new theory that expressions from the older theory can be logically derived from the new theory under certain conditions. The main function of this deductive use of the principle is, according to Fadner, that in this way the plausibility of the new theory will be considerably raised. In an inductive application, we encounter a genuine heuristic strategy: in this case, the form and/or content of the new theory is suggested by or inferred from that of its predecessor.

Fadner stresses two important issues regarding these definitions. Like Krajewski, he points out that in general we are not concerned with correspondence between entire theories but only between certain of their "operational equations." Next, Fadner emphasizes that this correspondence is only significant if both equations are also empirically adequate in the domain in question. This entails that the correspondence between operational equations is always "within a certain accuracy." First, because of the inevitability of measuring errors and of the finiteness of the number of data; second, because of the fact that in many cases the experimental data do not refer to the limit itself but only to the domain in the neighborhood of the limit. That is, in most cases, the predicted values will not agree exactly, though they should lie within a previously fixed interval. Summarizing, Fadner defines the generalized correspondence principle as follows:

> The operational equations of any new theory must reduce, within the appropriate accuracy, to the corresponding operational equations of a well-established previous theory in the "regions" where the previous theory is well supported by data. (Fadner, 1985, 832)

He adds to this that not only the mathematical equations but also the analogous terms from these equations should correspond to each other. Such a "term correspondence" implies that the terms in question have the same operational meaning. This applies for instance to the Newtonian mass m_0 and the relativistic mass m in the domain where the velocity v is much smaller than the velocity of light c. According to Fadner, such a continuity of operational meaning is not generally valid but restricted to some specific domains.

3.3 THE CORRESPONDENCE PRINCIPLE AND THE RISE OF QUANTUM MECHANICS, 1913–1925

In sections 3.5 and 3.6 I will analyze and evaluate these four views on the generalized correspondence principle in more detail. For this purpose I will first examine the role of the correspondence principle in physics and explain how it has been and still is used in concrete scientific practice. In the present section, the various parts played by the correspondence principle during the rise of modern quantum mechanics will be summarized.[3]

This case study mainly deals with developments within atomic physics in the years 1913 through 1925. Three kinds of theories are involved: the classical ones (mechanics and electrodynamics), Bohr's atomic theory, and modern quantum theory in the form of Heisenberg's matrix mechanics. I will focus in particular on the work of Bohr, Sommerfeld, Kramers, Born, and Heisenberg. In analyzing the history of the correspondence principle, we can discern three successive phases, in which the principle was used in different ways. This will be seen to be of great importance considering the philosophical argument.

1913–1915: Numerical Correspondence

The first phase ranges from 1913 to about 1915 and concerns correspondence as a numerical agreement of the values of some physical quantities in classical mechanics and electrodynamics and in Bohr's atomic theory. Bohr's theory can be characterized by the following two assumptions. First, the electrons in an atom are allowed to revolve only in well-defined, discrete (circular or elliptical) orbits around the atomic nucleus, each of these so-called stationary states corresponding to a definite value of the electron energy. Second, emission of radiation only takes place when an electron jumps from a stationary state with a higher energy to one with a lower energy. Both assumptions are definitely contradictory to what is expected on the basis of the classical theories.

Bohr examined a certain domain of phenomena, namely the frequencies of series of spectral lines of light emitted by atoms (Bohr, 1913; cf. Heilbron and Kuhn, 1969). In the first instance, he dealt with the hydrogen atom. On the basis of his atomic model, he was led to expect that the *numerical values* of the classical frequency $\omega_{n,\tau}$ ($= \tau \omega_n$, by definition) and of the quantum frequency $v_n^{n-\tau}$ would be approximately identical for a certain part of these spectra, namely:

$$v_n^{n-\tau} = \omega_{n,\tau} \text{ for } n \gg 1 \text{ and } \tau \ll n \qquad (1)$$

In other words, in the domain of large quantum numbers n, the quantum frequency of the radiation emitted in an electron transition from state n to state n-τ ($\tau = 1,2, \ldots$) agrees with the classical radiation frequency $\omega_{n,\tau}$, which, according to classical electrodynamics, equals the τ-th harmonic of the classical frequency of motion ω_n of an electron in the orbit n. Bohr tested this correspondence hypothesis by using it as a premise in a theoretical calculation of the Rydberg constant. Indeed, the calculated value of the constant appeared to agree well with the experimentally determined value.

We should notice that Bohr always speaks of "agreement of calculations" concerning this type of correspondence (or analogy, as he used to call it in those years). He himself was well aware of, and also plainly pointed out, the fact that despite this numerical agreement, at a theoretical level a large conceptual gap existed between his atomic theory and classical mechanics and electrodynamics.

1916–1922: Correspondence and Conceptual Continuity

In the second phase, however, it was assumed that in due course this conceptual gap could be bridged.[4] Two advancements may account for this modified assessment. First, Sommerfeld's quantum conditions and their further elaborations made it possible to provide a mathematical-theoretical proof of the correspondence between the frequencies (equation 1), not only in the case of hydrogen but, at least in principle, for all atoms that could be described as "multiply periodic systems." At the same time, the number of applications of Bohr's atomic theory strongly increased. Thus, among other things, successful explanations were proposed of the fine structure of the hydrogen spectra and of the so-called Stark effect.

Second, in this phase Bohr succeeded in incorporating not only the frequencies but also the intensities and polarizations of spectral lines into his theory. In doing so, he again made use of a correspondence argument. If we know the quantum theoretical transition probability $A_n^{n-\tau}$, we will be able to calculate, with the help of supplementary assumptions, the intensity $I_n^{n-\tau}$ of the spectral line originating from an electron transition between state n and state n-τ. Bohr, in the first instance, suggested that the following numerical correspondence should hold in this case:

$$A_n^{n-\tau} = f(C_{n,\tau}, \omega_{n,\tau}) \text{ for } n \gg 1 \text{ and } \tau \ll n \qquad (2)$$

(See Bohr, 1918; Kramers, 1919. $C_{n,\tau}$ are the Fourier components of the generalized position coordinates, which determine the form and the size of the orbit; f is a function known from classical electrodynamics.)

But in this second phase, Bohr moved considerably beyond such a numerical correspondence restricted to a specific domain. He postulated that a correspondence such as equation 2

> may clearly be expected to be of a general nature. Although, of course, we cannot without a detailed theory of the mechanism of transition obtain an exact calculation of the . . . probabilities, unless n is large, we may expect that also for small values of n the amplitude of the harmonic vibrations . . . will in some way give a measure for the probability of a transition. (Bohr, 1918, 110; see also 130–131)

Kramers developed this idea of a correspondence valid for all quantum numbers in more detail (1919, 47–50). The quantum frequency $v_n^{n-\tau}$ can be written as a certain average $<\omega_{n,\tau}>$ for all classically permitted orbits with energies lying between E_n and $E_{n-\tau}$. That is the reason why Kramers assumes that the following general correspondence is valid, too:

$$A_n^{n-\tau} = f(<C_{n,\tau}>,<\omega_{n,\tau}>) \text{ for all } n \text{ and } \tau \qquad (3)$$

Though there were problems in finding and carrying through the right way of averaging at the time, it was understood that these problems were merely of a technical nature and that they would be solved in the foreseeable future. This assumption was supported by the fact that Kramers's extensive calculations of two series of intensities of spectral lines showed a "suggestive" and a "convincing" agreement with the experimental data, also for smaller quantum numbers.

Altogether, then, in this second phase, the attitude towards the prospects of Bohr's theory was rather optimistic. As Bohr himself wrote at the end of 1917:

> It seems now really possible to a certain degree to overlook the theory with all its different applications from a uniform point of view. . . . At the present I am myself most optimistic as regards the future of the theory.[5]

The critical point for us is that now, besides a numerical correspondence between classical and quantum theory, a *conceptual correspondence* is postulated through the *orbital* atomic model associated with equation 3; a correspondence that is valid moreover for *all* quantum numbers, that is, for the entire domain of atomic phenomena. The same fundamental concepts $\omega_{n,\tau}$ and $C_{n,\tau}$, which govern the motion of the electrons in their orbits, and the same function f determining the transition probabilities are claimed to underlie both kinds of theory.[6]

1923–1925: Numerical and Formal Correspondence

In the third phase, however, the optimism of the second phase turned out to be unjustified. Bohr's theory encountered increasingly more empirical and conceptual problems, in particular with respect to the energies and spectra of many electron atoms and the interactions between radiation and matter such as dispersion and fluorescence (see, e.g., Jammer, 1966, ch. 3 and 4). In 1925 these anomalies brought about the abandonment of Bohr's atomic theory in terms of electron orbits to be described by classical mechanics in favor of Heisenberg's matrix mechanics that, together with Schrödinger's wave mechanics, marked the beginnings of modern quantum theory. Ever since, Bohr's atomic theory is referred to as "the old quantum theory."

In the transition phase, between 1923 and 1925, correspondence arguments played again a vital, albeit an unmistakably altered, role. Now conceptual correspondence, linked as it was to the mechanical orbital model, was rejected as a basis for atomic theory, and the orbital model was provisionally replaced by the so-called virtual field model.[7] In this model, an atom in a certain stationary state is conceived in terms of a probability field containing transition amplitudes $C_n^{n-\tau}$ that determine the transition probabilities $A_n^{n-\tau}$ via the completely quantum theoretical formula:

$$A_n^{n-\tau} = f(C_n^{n-\tau}, v_n^{n-\tau}) \qquad (4)$$

In two ways Kramers, Born, and Heisenberg employed correspondence arguments in applying this radically unclassical virtual field model. First, analogous to equation 1, they used a numerical correspondence for large quantum numbers:

$$C_n^{n-\tau} = C_{n,\tau} \text{ for } n \gg 1 \text{ and } \tau \ll n \qquad (5)$$

Second, Born (1924) generalized the procedure by which Kramers arrived at his quantum theory of dispersion. Born obtained an explicit rule for transforming the derivatives of classical quantities $\Phi_{n,\tau}$ to the action coordinates J (which frequently occurred in mathematical calculations concerning physical problems in the atomic domain) into the correct quantum theoretical expressions:

$$\tau(\partial \Phi_{n,\tau}/\partial J) \Rightarrow (\Phi_{n+\tau}^n - \Phi_n^{n-\tau})/h \text{ for all } n \text{ and } \tau \qquad (6)$$

Born's quantization rule (6), implying the systematic replacement of classical differentials with quantum differences, was one of the cornerstones of Heisenberg's matrix mechanics. This heuristic rule was distilled from a formal intertheoretical relation, in which left hand and right hand side originate from theories that are radically different from a conceptual point of view, as a consequence of the diverging concepts of the orbital and the virtual field model of the atom. On this basis we may also construct, in analogy to the second phase, a term correspondence holding for all quantum numbers (e.g., between $C_{n,\tau}$ and $C_n^{n-\tau}$ or between $\omega_{n,\tau}$ and $v_n^{n-\tau}$). But, just like Born's intertheoretical relation and his quantization rule, these relations between terms are *formal* (not conceptual) *correspondences*, expressing the existence of certain relations of mathemathical identity or substitution.

In section 3.2, we have seen that the generalized correspondence principle can be applied in two general ways: from the new theory L to the old theory S (Post, Krajewski, Fadner) and from the old theory S to the new theory L (Zahar, Fadner). Correspondence "from L to S" means that new theories L should satisfy constraints induced by S, while correspondence "from S to L" specifies a way for inferring (parts of) L from S. The examples of correspondence arguments during the rise of modern quantum mechanics discussed in this section can be classified in the same general manner. First, the numerical correspondences in the first and third phases imply a limit condition to be satisfied by L to the effect that for large quantum numbers, the theoretical properties $\Phi_n^{n-\tau}$ should quantitatively agree with their classical counterparts $\Phi_{n,\tau}$. In section 3.2, it was stated that such correspondences from L to S may considerably increase the plausibility of L. Thus it is a historical fact that Bohr's calculation of the Rydberg constant was perceived by his contemporaries as significantly supporting his atomic theory. Second, the formal correspondences in the third phase implied, via Born's quantization rule, a heuristic strategy for obtaining the correct form of the quantum mechanical equations from the corresponding classical ones. Thus, here we are dealing with a heuristic rule from S to L. This rule was applied successfully by Kramers in his dispersion theory and, more generally, by Heisenberg in constructing his matrix mechanics.

In surveying the whole episode in the history of twentieth-century physics, we may conclude that, ultimately, the success of the correspondence principle appears not to rest upon a conceptual correspondence but rather upon a combination of numerical and formal correspondence.

3.4 CORRESPONDENCE IN MODERN QUANTUM THEORY

Up to now I have examined the role of correspondence arguments during the genesis of quantum mechanics. However, also *within* modern quantum theory, the correspondence principle has proved to be an indispensable tool. In this case the principle, as an intertheoretical relation, is used not in constructing a new theory but in articulating and developing an existing theory. In this section I will merely sum up the forms and uses of correspondence in the context of normal quantum theoretical research. The question of the heuristic role of these correspondences will be dealt with later on, in section 3.6. Again, we come across the two ways of applying the principle that have been mentioned above. Let us first look at the correspondence from L to S. Here essentially three kinds of arguments can be found (Messiah, 1969, 214–241).

1a. The first argument aims at showing that, under certain conditions, quantum mechanical objects (approximately) behave like ordinary particles described by classical mechanics. The best known example of this is the Ehrenfest theorem. For a one-particle system subjected to a force $F(r)$, the expectation values $<r>$ and $<F(r)>$ satisfy the relation $md^2<r>/dt^2 = <F(r)>$. Then, under the conditions that $<F(r)> = F(<r>)$ and that the position wave packet is and remains sharply peaked, we (approximately) have: $md^2r/dt^2 = F(r)$, Newton's second law. Thus, if these conditions apply, our quantum mechanical object can be described as a classical particle. In other words, in this case a *conceptual* continuity seems to obtain between the classical and the quantum mechanical descriptions. A correspondence has been constructed that is not merely based on a limiting process between formulas but that is of a conceptual nature. The occurrence of such correspondences, however, is exceptional. In fact the above-mentioned conditions do apply only to rather special cases, namely to a limited class of force functions or during very restricted intervals of time (cf. Messiah, 1969, 216–222).

1b. The second procedure to bring about a correspondence transformation from modern quantum mechanics into classical mechanics is more general (see Messiah, 1969, 222–224). It is also possible to formally relate Newton's second law to the quantum mechanical Schrödinger equation by taking the *mathematical* limit $h \to 0$.[8] The interpretation of both equations, however, remains completely different. For instance, in one of the possible approaches (the "hydrodynamical analogy") *one* quantum mechanical object described by the Schrödinger equation corresponds to a fluid *current* of noninteracting classical particles to be described by Newton's second law.[9] Consequently, in these cases the correspondence is obviously not of a conceptual but rather of a *formal* nature; it holds good for the mathematical formulas in question.

1c. Third, we may also try to construct *numerical* correspondences by *physically* specifying limit domains corresponding to $h \to 0$. It is possible to show, for instance, that for periodic systems in many cases (but not always: see Liboff, 1984), taking the limit $h \to 0$ is equivalent to a restriction to the domain of large quantum numbers ($n \to \infty$). We can derive numerical correspondences in that domain, e.g., concerning the energies or frequencies of the hydrogen atom, by taking the simultaneous limits of $h \to 0$ and $n \to \infty$, under the constraint that nh equals the classical action of the system in question (Hassoun and Kobe, 1989). Another example is supplied by Coulomb scattering, for instance the scattering of a proton by an atomic nucleus. Here $h \to 0$ corresponds to the domain of small scattering angles. In that domain, a numerical correspondence is known to exist between the classical and the quantum mechanical value of the differential cross section (Messiah, 1969, 228–231).

2. In the second place the correspondence principle is also used to argue from S to L, that is, as a rule for transforming classical equations into quantum mechanical equations. This rule is quite general, stating that we should formally replace classical quantities and functions of them with corresponding operators and functions of them in a Hilbert space, and that the numerical values of the classical quantities correspond to the eigenvalues of the eigenvectors of the corresponding operators. This procedure is clearly analogous to Born's rule of substituting differentials by differences. And just as Born's rule, it does not bring about a conceptual correspondence.

This general quantization rule is used frequently in modern quantum theory and also in quantum field theory. In the context of our argumentation, two further observations should be made. First of all, generally speaking the rule does not produce correct quantum mechanical results all by itself. Supplementary assumptions are necessary too, for instance concerning the choice of a suitable type of coordinates or regarding the precise mathematical form in which to express the classical functions (see Messiah, 1969, 68–71). These assumptions may be made plausible either by referring to the fact that they lead to successful experimental predictions or on the basis of theoretical considerations, e.g., requirements of symmetrization. Second, we should observe that this correspondence from S to L does not produce all of quantum mechanics. It will give you only those quantum mechanical quantities and expressions for which a classical analogon exists. Irreducibly quantum mechanical properties, such as parity, or principles, like Pauli's exclusion principle, will necessarily remain outside the scope of this procedure.

3.5 EVALUATION OF THE GENERALIZED CORRESPONDENCE PRINCIPLE

On the basis of the material collected in the previous two sections I will now proceed to evaluate the views on the generalized correspondence principle summarized in section 3.2. Although so far I have taken into account mainly quantum theory, it turned out that even in this single case many and diverse forms of correspondence occur, so that an evaluation based upon this certainly seems sensible. Moreover, whenever appropriate, I will also discuss other examples of correspondence. The issue will be approached via two questions: what is the *nature* of the correspondence relation; and *what* does this relation really refer to?

The Nature of the Correspondence Relation

We have already seen that it is important to distinguish three forms of correspondence: numerical correspondence between the values of physical quantities from different theories; conceptual correspondence between theoretical terms; and formal correspondence between mathematical symbols and expressions. The most significant continuities between theories appeared to be the numerical and formal correspondences, whereas conceptual continuity is rather the exception than the rule. Surely, the latter not only obtains in the case of relations between classical and quantum physics. Following the work of Hanson, Kuhn, and Feyerabend, many historians and philosophers have shown that the occurrence of conceptual discontinuities is a normal aspect of intertheory relations in the natural sciences (see, e.g., Laudan, 1981, esp. 32–45).

Generally speaking, the advocates of a generalized correspondence principle have, to a greater or lesser extent, acknowledged the significance of conceptual breaks. Zahar is the most outspoken:

> In all these cases we have to do with a syntactic-mathematical type of continuity which, together with the semantic stability of our observation language, entails continuity at the *empirical* level. This is consistent with the occurrence of dramatic revolutions at the ontological level. (Zahar, 1983, 247)

Post's position on this point is more ambiguous. He postulates that the L theory will often embrace a considerable part of the lower theoretical structure of the S theory (Post, 1971, 229 and 231). If we interpret this (as Laudan, 1981, 37–42, does) as implying that there will always be a continuity with respect to the conceptual-theoretical aspects of successive theories, then it is obviously wrong. Often new theories imply a reinterpretation *also* of the lower theoretical laws. A well-known example from modern quantum theory is the wave-particle duality that leads to a radically different conceptual interpretation of light and its properties and, thus, of the lower theoretical spectral laws.

It is also possible to view the theoretical continuity claimed by Post as a formal correspondence. The cited example of the conservation of the ordering of the elements in the periodic table, however they and their properties might be conceptually reinterpreted, as well as some other passages in Post's 1971 paper (e.g., 237–238) give cause for such a view. In this case, Laudan's criticism, based as it is on his (incorrect) assertion "that T_1 can be a limiting case of T_2 only if *all* the entities postulated by T_1 occur in the ontology of T_2," is not justified.[10] And the same applies to Koertge's interpretation, which says that the generalized correspondence principle "tends to underemphasize the radical changes which occur during scientific revolutions" (Koertge, 1973, 177).

Krajewski's view on the matter is not completely satisfactory either. It is especially his claim that the correspondence between classical mechanics and special relativity is homogeneous, or conceptual, that is questionable. According to him, this claim is justified since in both theories the term "mass" has the common meaning of "the measure of inertia." But what is inertia? It is the property of massive bodies to resist changes of their velocity. Classical and relativistic velocities, however, contrast considerably. They obey different addition and transformation laws. Even more importantly, they are defined on the basis of radically disparate notions of space and time. Thus, this argument boils down to the claim that a conceptual discontinuity between classical and relativistic mechanics does exist, though it does *not primarily* concern the notions of mass but rather the radically divergent conceptions of space and time and of space-time (cf. also Zahar, 1976, 235–237 and 247–248).

Let us now turn to numerical correspondence. Remarkably enough, this form of correspondence remains strongly underexposed, especially in the accounts by Post, Krajewski, and Zahar. Krajewski, for one, greatly emphasizes the formal-logical character of the correspondence relation, defining it at one place for laws L_1 and L_2 as a relation in which

> the equation $F_2(x) = 0$ of L_2 passes asymptotically into the equation $F_1(x) = 0$ of L_1 when some parameters p_i characteristic for L_2 tend to zero. (Krajewski, 1977, 42)

Such a formal correspondence between mathematical equations is by itself insufficient, however. After all, in applying the correspondence principle, we are not dealing with a relation between arbitrary theories but between theories that in part cover the same domain of phenomena in reality. Formal correspondence alone does not at all guarantee that this will be the case. The same mathematical equations may well describe completely different kinds of phenomena (cf. Feynman, Leighton, and Sands, 1964, ch. 12).

Because of this, numerical correspondence is required apart from formal correspondence. In other words, the mathematical limit (e.g., $h \to 0$) has to be physically specified for various domains (e.g., large quantum numbers for the spectra; small angles in scattering experiments), and the formally corresponding terms or quantities need to have the same experimentally obtainable values in (the neighborhood of) that limit.[11] We thus see that the applicability of the generalized correspondence principle to successive theories presupposes a certain kind of stability or continuity at the empirical level, especially when the theories in question are conceptually discontinuous (see Radder, 1988, chs. 3, 4, and 5).

The Relata of the Correspondence Relation

The second question to be answered in this section is: *what* is it that is claimed to be corresponding by the generalized correspondence principle (cf. also Bunge, 1970, 288–296)? A clear answer to this question is required if we want to establish the heuristic power of the principle. As we have seen, in particular Post presents the generalized correspondence principle mainly as a correspondence from L to S^*. However, it is not true that under the right conditions L *entirely* transforms into *all* of S^*.

On the one hand, by no means all propositions or equations from L lead to statements or expressions in S^*. To mention two examples: the expression for the rest energy in the theory of special relativity, $E = m_0c^2$, does not contain v and consequently taking the low velocity limit is pointless. An illustration from quantum mechanics is the statement that the state of a system of N identical particles is either symmetrical or antisymmetrical under permutation of the particles. There is no limit condition implying the transformation of this statement into a statement from classical mechanics. In view of the fact that a good new theory will also explain and predict completely novel phenomena and not merely replicate its predecessors, all this is of course not surprising.

Fadner also enters into this question. He argues that correspondence is not a relation between entire theories but merely between some specific "operational equations." The Schrödinger equation in quantum mechanics, the second law in Newtonian mechanics, and the expression for the total energy, $E = mc^2$, in Einstein's theory of special relativity are instances of such operational equations. This specification is not clarifying, however, mainly because of the fact that his use of the term "operational" remains rather obscure. After all, the previously mentioned examples are in fact far from being operational equations "from which deductions about the values of observable quantities can be made quite directly" (Fadner, 1985, 831). On the contrary, such fundamental theoretical equations have a long way to go before they can be empirically tested, as has been shown convincingly by Cartwright (1983). Moreover, not all of the examples are to the point. For instance, the total relativistic energy of a particle $E = mc^2$ transforms into $E = m_0c^2$ (for $v/c \ll 1$), which is not a valid operational equation within the classical theory. In conclusion, it is more adequate to claim that formal correspondences may obtain between *some* expressions, which are not themselves operational but which may be further articulated towards various empirical domains in order to test them also for numerical correspondence.

On the other hand, there was the question of whether the claimed correspondence from L to S^* gives us *all* of S^*. Post, in particular, answers this question definitely in the affirmative: according to him, the phenomenon of "Kuhn loss" simply does not exist (e.g., Post, 1971, 229). As we have already seen, this is certainly not true in the case of conceptual correspondence. Moreover, the idea that successive theories in the history of science are empirically ever more successful, i.e., that later theories retain the complete empirical success of earlier ones, is a myth. As we have seen, numerical correspondence refers only to specific domains. And although, of course, there are fields in which the new theory is empirically successful while the old theory fails, it is equally true that domains exist where the old theory remains required for solving empirical problems and where the new theory, at best, applies only "in principle." Take the phenomena of fluid flow. In this domain, it is actually impossible to derive, for example, Poiseuille's law (which relates the rate of flow to the pressure, the internal friction, and the dimensions of the flow tube) on the basis of a purely quantum mechanical treatment of the fluid molecules. For phenomena like these, quantum physics is practically useless and classical physics obviously superior. In conclusion, Post, in his rejection of Kuhn loss, wrongly infers from the possibility of formally explaining why S^* worked (by means of a mathematical correspondence between S^* and L) that L and S^* will possess the same concrete puzzle-solving capabilities.

Finally, I will make a few remarks concerning Zahar's use of the generalized correspondence principle from S^* to L. In view of everything said so far, I can be rather brief about this. For a start, in applying the principle in this way, we do not make use of S^* or parts of it alone. Zahar (1983, 248–252) realizes this. One example was given in section 3.4: generally, certain symmetrization requirements and/or empirical arguments should be added before we are able to quantize the classical formulas in the right manner. Furthermore, Zahar's procedure will not give us the entire L theory. As we have seen, certain parts of L will not have an analogon in S. Hence, these parts (e.g., the exclusion principle in quantum mechanics) cannot be derived on the basis of a correspondence to S^*.

On the ground of the previous discussion, I can *conclude* that, roughly, the proponents of a generalized correspondence principle are right: the principle plays an important role in linking up successive theories, at least in the more mathematicized, physical sciences.[12] Nevertheless, it will also be clear that the principle has to be qualified with respect to its nature and that it has to be relativized with respect to its range of application. In general the correspondence is not of a conceptual but rather of a formal and numerical nature. And it does, moreover, not apply to entire theories. The formal correspondence does not relate all terms and expressions of L and S^*; and the numerical correspondence, not all physical quantities.

It will be obvious that these qualifications and relativizations are crucially important in establishing the heuristic scope of the generalized correspondence principle. Consider, for instance, Kramers's failure to achieve a decisive breakthrough in quantum electrodynamics (in the 1930s and 1940s). Dresden blames this failure on Kramers's unwarranted belief in the universal validity and unrestricted applicability of the correspondence principle:

> Crucial to Kramers was the existence of correspondence principles which guaranteed the possibility of smooth, continuous transitions from the relativistic as well as the quantum domains to the purely classical regimes. . . . It was this unwillingness, or inability, to detach himself from classical visualizable physics or to make a decisive, irreversible break with classical concepts that ultimately defined the limitations which he imposed on himself and his work. (Dresden, 1987, 429)

Of course all this does not imply that an unqualified and unrestricted use of the correspondence principle will necessarily fail in every case. But it does entail that *when generalizing* the principle, we should take into account the mentioned qualifications and relativizations.

3.6 CORRESPONDENCE AND HEURISTICS

It is important to distinguish between philosophical reconstructions of relations between theories in terms of correspondence and the actual heuristic use of the principle by scientists. The fact that the former is possible in a certain case does not imply that the latter actually occurred. Therefore, the account of the genesis of matrix mechanics in section 3.3 is most relevant to the question of the actual heuristic role of the correspondence principle. The discussion of its normal-scientific use, in section 3.4, requires some explanation in the present context.

In the first place, there was the rule of replacing classical quantities with quantum operators. This rule is widely applied in quantum mechanics and field theory, and it is undeniably of great heuristic value. Second, the classical approximation methods (by means of $h \to 0$) are also used heuristically, be it mainly in tackling more specific problems, for instance in probing a problem or for obtaining approximate results when a more rigorous solution is not feasible. In contrast, the third form of correspondence, by means of the Ehrenfest theorem, has a limited heuristic value. But after what has been said so far about conceptual correspondence, this will not come as a great surprise. Thus, in contrast to Post (1971, 233), we see that in the case of the relation between classical and quantum mechanics, a number of correspondences did and do play a heuristic role, even though the Ehrenfest theorem is not one of them (see Zahar, 1976, 1983, for other examples).

What, then, is the relationship between the generalized correspondence principle and heuristics, the rule-governed process of transforming knowledge? *At first sight* the correspondence principle, at least in the formulations of Post and Krajewski, seems to be only heuristically relevant in a weaker sense. The principle is, after all, claimed to be mainly "conservative," in the sense of retaining the old. More precisely, in this form it strengthens and specifies the old theory: via the correspondence between L and S^*, we have traced the limits of validity of S (cf. also Rohrlich and Hardin, 1983). Post states it as follows:

> The catch in using the correspondence principle heuristically is of course that the true extent of S^* is only conjecture at any one time. (Post, 1971, 231)

But if this is the case, if the generalized correspondence principle merely teaches us something about the limitations of the old theory S and nothing about the development and the validity of a new theory L *outside* the domain of overlap with S, why do we use the principle at all? Why do we not just study S, independently of L, in order to obtain S^* and search for a new theory L, independently of S, that is adequate on domains where S fails?

To answer these questions, we should first of all bear in mind the fact that the generalized correspondence principle applies precisely to theories that cover at least partly the same set of phenomena: a point emphasized already in the preceding sections. Since S was partly successful in handling these phenomena, it has to contain "somewhere," "in one way or another," or "partially" useful insights. Therefore it seems reasonable to somehow make use of these insights when constructing a new theory L. At this point it remains an open question what these qualifications of "somewhere," etc. are meant to imply. Many philosophers, however, have attempted to specify and explain this intertheoretical continuity by means of, in particular, *realist* philosophical theories. I will come back to this issue in the next section.

Let us now turn directly to the question of the heuristic power of the generalized correspondence principle. With respect to this question, we must realize that the principle is not used exclusively in the above sense, from L to S^*. It is also used to reason from S to L, a point stressed in particular by Zahar (and to a lesser extent by Fadner). In sections 3.3 and 3.4, we have seen a number of examples of this use.

Born's quantization rule, discussed in section 3.3, is the most explicit. It first tries to bridge the conceptual and numerical gap between S and L by means of a formal framework that is, in principle, much wider applicable than only to the domain of the (numerical) correspondence conditions. Born's replacement of classical differentials with quantum theoretical differences not only brought about numerical correspondences in the domain of overlap with the classical theories, it also constituted one of the major steps towards a new theory: namely, Heisenberg's matrix mechanics. Of course such bridging attempts remain fallible, but in trying to generalize the successful formal aspects of the old theory S, they are in any case based upon some firm points of support on the "old" side of the gap.[13]

Yet, the construction of a formal framework is clearly insufficient for landing on the right part of the opposite side of the gap. Formal correspondence is not enough: in order to be an improvement, the new theory L has to be also empirically adequate in some domains in which S failed. However, in order to judge the empirical adequacy of a set of formal equations, a conceptual interpretation or model is required—perhaps not in a logical sense but certainly in a practical sense. In the case of Born's formal correspondence, this role was played by the virtual field model mentioned in section 3.3. Its main characteristics were that all atomic quantities depend essentially on two discrete stationary states (and not on one electron orbit, as was the case in Bohr's mechanical model) and that the virtual field is a probability field, determining not the transitions themselves but only the transition probabilities. In this way, this conceptual model caught the three major distinguishing features of modern quantum theory, namely its discrete or quantum character, its possibility of superpositions between states, and its probabilistic features.

It is these conceptual innovations that most clearly exhibit discontinuity and a creative switch. This raises the question of the grounds on which such new conceptual interpretations or models are held to be plausible at all, especially in a first stage in which they at best imply a promise of empirical success. It seems to me that in general such plausibility will be strongly context dependent. Consider for instance the virtual field model of the atom discussed above. Elsewhere (Radder, 1983) I have explained, by reformulating Forman's 1971 analyses, that at least in the case of Kramers but probably also in that of Born, Bohr, Pauli, and Heisenberg, the model derived its plausibility partly from the failure of Bohr's orbital model and partly from its accompanying "epistemological positivism," which was quite generally accepted among the leading physicists of those days. This kind of positivism, in its turn, was closely linked to what Forman (1971) calls the "intellectual climate of the Weimar Republic."

The "metaphysical principle" (see Zahar, 1983, 248–249) behind this epistemological positivism was that science can never discover the essence of the world and that, therefore, theories are no more than formal schemes, which are adequate if they lead to a correct description of the empirical data. For this reason, there is no need at all to stick to Bohr's realistically interpreted, orbital model. Instead the virtual field model is much more preferable, so it was argued, because it explicitly starts from such an epistemological positivist point of view (Radder, 1983, 172–176). This example of the contextual construction of the plausibility of newly proposed theories can be easily supplemented by many other case studies that the sociology of scientific knowledge has produced by now.[14] As a consequence of these contingent conceptual aspects, the heuristics of scientific knowledge transformations is and is bound to be an open process. It can never be caught in universally valid heuristic rules.

The argument so far implies a criticism of Zahar's view of a rational deductivist heuristics. According to him, heuristics not only employs scientific premises (such as the correspondence principle) but also stable metaphysical principles, which are rooted in daily life (and probably even innate or genetically determined) and which are being specified and developed for scientific use in the positive heuristic of research programs. The claim of stability or constancy of these principles seems, however, simply wrong. In antiquity and the Middle Ages, for instance, people did not assume the uniformity of nature as a principle but, on the contrary, viewed nature as constituted by qualitatively different domains, each with its proper kind of phenomena and laws. The above-mentioned principle that science can never discover the essence of the world, is not universally agreed upon either. Moreover, Zahar claims that in concrete cases, a coherent choice should be made from all available metaphysical principles. Again we may ask: from what does any particular choice (e.g., of a probabilistic instead of a mechanistic world view) derive its plausibility?

The same arguments make it possible to correct Zahar's claim of the deductivity of heuristic reasoning.[15] More precisely, they show that deducibility is only one of the smaller issues. After all, it does not take great pains to make almost any argumentation (whether it is scientific or not) deductive, if only we are allowed to add any premises we like. Of course, what matters is the status of the premises added. Hence, the central problems of heuristics (and of philosophy of science in general) do not primarily concern the deductivity of argumentations but rather the plausibility of their premises.[16] In other words, the crucial point is not so much the deductivity of a certain argument but the formulation of *precisely this* (deductive) argument that offers an acceptable new theory because of the very fact that its premises are plausible.

3.7 PHILOSOPHICAL CONCLUSIONS

Up to now I have been concerned primarily with expounding and evaluating (views on) the actual use of the generalized correspondence principle. In this concluding section, I will briefly deal with three philosophical issues that are directly related to the above discussions: the relation between correspondence and theory reduction, the question of why applying the principle is justified, and the issue of the discovery-justification distinction.

At the beginning of the last section, I distinguished reconstructed and heuristic intertheory relations and subsequently concentrated on the latter. Yet, the present considerations are also relevant to the former. For instance, the correspondence from L to S^* may be viewed as a form of *reduction* if its heuristic use has been successful, so that it has led to more or less stable theories L and S^*. I will now briefly point out some of the consequences of my analyses with respect to the debates about reduction.

A starting point for many discussions forms Nagel (1961, ch. 11). In his view, S^* reduces to L if the entire theory S^* can be logically derived from L. In contrast, early critics (such as Feyerabend, 1962; Kuhn, 1970b) stressed the incommensurability of S and L, considered as self-contained wholes, and hence rejected the possibility of reduction completely. More recent approaches tend to avoid both extremes. They point to the fact that different forms of reduction, requiring different philosophical treatments, occur (Nickles, 1973, 1977); or that reduction pertains exclusively to certain aspects of S and L (Rohrlich, 1988). The present analysis is, broadly speaking, in line with these more differentiated approaches.

Yet, these approaches are not completely satisfactory, because they generally require even more differentiation. Nickles, for instance, views the Ehrenfest theorem as an important example of one type of reduction. But his account of the theorem, saying that "*any* relation that appears in classical mechanics must be valid as a relation between quantum theoretical expectation values" (1973, 194, emphasis added), is not correct. For, as we have seen in section 3.4, the Ehrenfest theorem applies only to rather special cases. Furthermore, in his 1977 article (p. 587), Nickles endorses Post's claim of no empirical Kuhn loss for the case of "domain-preserving reductions." Obviously, the criticism of this claim given in section 3.5 applies to this case as well.

As a further example, consider Rohrlich's (1988) view that reduction exclusively concerns the mathematical frameworks of S and L and not their conceptual interpretations. If a reduction relation holds, the mathematical structure of S can be rigorously deduced from that of L. In a mathematical sense, there is no Kuhn loss, according to Rohrlich. He furthermore claims that in the case of classical and quantum mechanics, it is the correspondence principle that makes such a mathematical reduction possible. On the basis of the analyses in this chapter, two observations are pertinent with respect to this view.

First, the claim that under the appropriate conditions the complete mathematical framework of classical mechanics can be rigorously derived from that of quantum mechanics does not seem to be valid. The basic problem is that classical observables are mathematically represented by *functions* (on a phase space of generalized coordinates), while quantum observables are represented by *operators* (on a Hilbert space of state vectors). Since functions and operators are different mathematical entities, it is not possible, even when h is small, to derive an equation of functions from an operator equation. For instance, the fundamental equation of motion for a classical observable—$dA_{cl}/dt = \{A_{cl}, H_{cl}\} + \partial A_{cl}/\partial t$—can only be mathematically *derived* if we, incorrectly, identify the classical functions A_{cl} and H_{cl} with the corresponding quantum mechanical operators A_{qu} and H_{qu} (cf. Messiah, 1969, 317–318). It is true, as pointed out in section 3.4, point 2, that a formal correspondence does exist in this case, in the sense that the classical and the quantum equations of motion have the same mathematical form. However, contrary to Rohrlich's claim, this formal correspondence does not entail that all classical formulas can be "rigorously deduced" from their quantum mechanical counterparts.

Second, in the case of a reduction (or correspondence) relation, we require that the theories in question in some way or other cover a common domain of reality. Rohrlich speaks of "finer" and "coarser" theories and requires that the domain of the coarser theory is completely contained within that of finer theory. Apart from the empirical Kuhn loss discussed in section 3.5, there is the problem that mathematical reduction alone cannot guarantee this. In order to meet this requirement, a notion of numerical correspondence should be taken into account too.

Next I will enter into a question that was touched upon in the preceding sections but not yet dealt with in sufficient detail. This is the question of *why* applying the principle is advisable and fruitful. One suggestion (cf. Nickles, 1985, 184–193) might be that a good reason for using the principle is that it helps us to generate new knowledge in an *efficient* manner. Efficiency, it might be argued, constitutes an essential part of any scientific methodology and consequently employing the correspondence principle is strongly preferable to the blind generation of new hypotheses and theories. By itself, however, this answer is incomplete and remains uninformative as long as it is not further explained *in what sense* the principle is efficient. After all, efficiency always means "efficiency with respect to a certain goal." So, it needs to be clarified which goal can be reached more efficiently with the help of the generalized correspondence principle than without it. At this point often realist philosophical interpretations are proposed.

A frequently occurring view is convergent realism. In the present context, this can for instance be found in Post's paper (1971, 240), although he does not explicitly urge it as a ground of the success of the correspondence principle. The general idea is that, even though scientific theories may never be true in an absolute sense, yet the successive theories of "mature" science approximate the truth increasingly better (Putnam, 1978, 19–22). Such convergence, however, requires a measure of long-term conceptual and ontological continuity which in general the history of science does not warrant. Especially when we add cases of conceptually discontinuous transitions within undeniably mature sciences, like the one discussed in section 3.3, Laudan's (1981) confutation of convergent realism is convincing.

Krajewski frames his views on the correspondence relation also within a philosophical theory, namely, the theory of idealization and factualization (Krajewski, 1977, chs. 2 and 8). This theory, which may be considered as a dialectical version of convergent realism, is based on a distinction between essence and appearance. Theoretical laws and theories are always (counterfactual) idealizations describing the essence of nature. Empirical tests and applications then demand a procedure of factualization (or concretization) in which the counterfactual assumptions are one by one replaced with empirically adequate statements. The correspondence relation is also interpreted in these terms by Krajewski: the counterfactual conditions $p_i = 0$ of the old theory S are transformed into the empirically correct assumptions $p_i \neq 0$ in the new theory L. That is to say, S is a more idealized theory and L a more factualized theory. It follows from this, however, that S (e.g., classical mechanics or Boyle's gas law) is a better approximation of the essence of the phenomena than L (quantum mechanics or Van der Waals's law, respectively). Krajewski (1977, 52) himself notices this problem, speaking of an "interesting paradox." This is really an understatement: in fact the problem seems to be a rather disastrous consequence of the philosophical views underlying the methodology of idealization and factualization.[17]

Altogether, then, it seems that forms of convergent realism are not capable of "explaining" the success of the generalized correspondence principle. In the preceding section, I stated that this success implies, first, that S and L, in part, refer to the *same* domains of phenomena and, second, that S in one way or another contains useful formal insights, which may be exploited in heuristically constructing L. If it is correct that it is impossible to elucidate this state of affairs by means of convergent realism, it is natural to look for a more moderate realist interpretation.

Elsewhere I have argued extensively how such a moderate realist interpretation of natural science may be constructed (Radder, 1988, part II; see also chapter 4). Its essence is that the theoretical terms of successful scientific theories do refer to elements of a human-independent reality and that formally corresponding terms from successive theories may co-refer to elements in their common domains. In contrast to convergent realism, all notions of conceptual representation, which always presuppose long-term conceptual continuities, are explicitly avoided in this "referential" realism. Experimental and mathematical nonlocalities are used, however. They allow for the constructing of numerical and formal correspondences and for incorporating them into a criterion for establishing coreference of differing theoretical terms (e.g., $\omega_{n,\tau}$ and $v_n^{n-\tau}$) to the same element in a common domain of phenomena.

In this way the use of the generalized correspondence principle may be interpreted in a realist sense. Its success is based on the fact that in the relevant domains, S and L deal with the same elements of reality and that the construction of formal intertheoretical relations and heuristic rules based on them constitutes an attempt at going beyond the old theory through a systematic procedure of generalization of some of its successful formal aspects.

Finally, I will briefly examine which conclusions can be drawn from the above treatment of the generalized correspondence principle with respect to the discovery-justification debate (cf. Nickles, 1980). The role of the correspondence principle in the genesis of quantum mechanics is particularly illuminating with respect to the nature of discovery. The account presented in section 3.3 makes it plain that this discovery is not at all adequately described as the result of an instantaneous flash of intuition but rather as a complex and prolonged process, in which several factors eventually produced a major innovation. Moreover, it was not necessary to have a look inside the heads of the discoverers in order to discern a coherent pattern of discovery. That is, discovery does not so much concern purely subjective thoughts of individuals but rather the production of intersubjective, scientifically meaningful problems and solutions.[18]

A much-debated issue is the relation between discovery and justification. What follows from our discussion of the generalized correspondence principle with respect to this issue? If one of the (many) aims of science is to generate new knowledge of the same objects in new domains (outside the range of applicability of the older theories), then using the principle in the way outlined above is one step in the right direction. As we have seen, the principle may be further developed into a condition for the coreference of theoretical terms. Since realism is one of the standard methods for justifying scientific knowledge, it may be argued that we are dealing here with an intrinsic, as opposed to a de facto, connection between discovery and justification (see Laudan, 1980; Nickles, 1985). However, this does not mean that discovery methods possess justificatory force *just because* they are used in constructing new knowledge. On the contrary, discovery by itself always presupposes some form of justification. It at least partially anticipates requirements it will have to meet when it is evaluated more systematically (see Gutting, 1980). This becomes obvious at once when we ask ourselves what a completely unjustified discovery would be a discovery of. Yet, it will be clear that the justifications at issue are those presented by the scientists themselves. And these do not infrequently differ from those recommended by philosophers. This important point is all too often overlooked in philosophical debates about discovery and justification.

For this and several other reasons, the discovery-justification distinction should not be a basic, let alone a defining, distinction for philosophy of science. Following my introductory remarks, I would propose to abandon discovery-justification talk in favor of a view of the development of scientific knowledge as a continuing process of transformation, which is essentially nonmonolithic and which may exhibit both nonlocal patterns and local idiosyncracies. In this process, argumentative support ("justification") plays a role at every stage, even though the evidence offered may deviate in different stages and even though the extent to which the evidence is accepted as supporting the claims may vary, either increase or decrease, in the course of the process.[19]

CHAPTER 4

Science, Realization, and Reality

4.1 CHANGE AND WORK

The analyses in the preceding chapters revealed a number of important dimensions of experimental and theoretical science. I will now make use of some of their results in dealing with the much-discussed philosophical issue of the relationship between science and reality. More specifically, I will tackle the question of whether or not true scientific knowledge is about an independent reality. Someone who is inclined to answer this question in the affirmative will have to face two different (though not unrelated) basic sources of difficulties concerning this answer. The first is "change," the second, "work." Let me explain.

If true knowledge means knowledge about a reality that is independent of the existence of human beings or of the process in which it has been produced, true knowledge is invariant knowledge. But, when we look at the historical development of actual knowledge claims endorsed by scientists or scientific communities, we often see radical changes rather than invariance. The representations of reality turn out to be variable both temporally and spatially. This fact entails a severe challenge to realist interpretations, if one rightfully wants to avoid one's interpretation to be completely disconnected from scientific practice. This "Kuhnian challenge" to scientific realism has been much debated in the history, philosophy, and sociology of science of the last decades. "Conceptual discontinuity," "incommensurability," and "interpretative flexibility" constitute the key notions of these debates.[1]

The second source of problems facing a realist interpretation of science is the work that is needed to produce scientific knowledge. Today virtually no one adheres any more to the view that knowledge is somehow imprinted by reality on passive knowers. Nearly all modern philosophers insist that acquiring scientific knowledge demands an active contribution from human beings. Because of this, antirealists may and do argue that the nature of the knowledge product is not at all defined by an independent reality but rather by the specific characteristics of this production process. The work done to produce and maintain a knowledge claim can, according to this antirealist view, fully account for its acceptance.

The problem "work" poses to realist interpretations becomes even bigger if, besides theorizing, we include and scrutinize experimenting as a substantial and specific part of scientific activity. In this case, we cannot restrict ourselves to taking account of linguistic and interpretative contributions, but we also have to deal with manipulative and material contributions. Clearly, these contributions are crucially significant, since most "natural" phenomena can survive only in the artificially produced environment of a laboratory.

Now, the fact that it is not merely the knowledge of these phenomena but also their existence that is essentially dependent on the specific, productive activities of the experimenters implies a strong challenge to realist interpretations of scientific knowledge claims. After all, given this fact, the descriptions of phenomena seem to refer to specific human products and not to an independent world; the theoretical explanations of these phenomena seem to explain specific human constructions; the experimental tests merely seem to test the extent to which theories are supported by the results of certain human practices; and finally, if scientific knowledge turns out to be valid outside the original laboratory, this seems to be simply due to the transference or re-creation of the laboratory conditions on a larger scale rather than to its being about an independent reality. I will call the challenge described above the "Bachelardian challenge," after the French philosopher who—already in the 1930s—emphasized the crucial significance of the productive character of scientific experimentation.[2] Simply stated, it is the question how scientific knowledge can be about a human-independent reality if this reality is so thoroughly dependent on human work.

In this chapter, I want to develop an adequate answer to the Bachelardian challenge. Accordingly, my focus will be on ontological issues. Yet, I think that such issues should not and cannot be dealt with independently of epistemological questions. For this reason I shall start, in section 4.2, with a brief summary of my proposals for meeting the Kuhnian challenge by means of a referentially realist epistemology. Next, section 4.3 begins the systematic discussion of the Bachelardian challenge. The basic claim is that the reproducibility of experiments entails an ontology of independently persisting, real potentialities and their historically contingent realizations. In section 4.4 this ontology is more substantially related to experimental practice on the basis of a further differentiation of the notion of experimental reproducibility. Section 4.5 develops some notable features of the proposed ontology by arguing for a (partial) rehabilitation of the role of "abstraction" in scientific practice. In section 4.6 two alternative views on the problems in question, taken respectively by Bhaskar and by Latour and Woolgar, are briefly discussed and evaluated. The final section deals with the question of the scope of the proposed approach. My conclusion is that the epistemological and ontological claims can be plausibly extended from the experimental to the (purely) observational sciences.

I started this introductory section by posing the question whether scientific knowledge can be about an independent reality. However, taking both the Kuhnian and the Bachelardian challenge seriously has led me to a certain transformation of this point of departure. That is to say, the main question at issue is rather this: is it feasible to develop a plausible realist epistemology and ontology for (experimental) science, starting from the premise that scientific knowledge is at once *in* and *about* the world?

4.2 MEETING THE KUHNIAN CHALLENGE: A REFERENTIALLY REALIST EPISTEMOLOGY FOR EXPERIMENTAL SCIENCE

The principal aim of this chapter is to propose a cogent reply to the Bachelardian challenge. In doing so, I will assume and use the results of some earlier work. In the present section I will briefly summarize those aspects that are needed for a proper understanding of what follows.

An important preliminary question is whether it is possible to meet the Kuhnian challenge. After all, if this is not the case, any attempt at meeting the Bachelardian challenge will be futile from the start. I think, however, that a modest form of epistemological realism, which is compatible with the phenomenon of recurrent sociohistorical change, can be defined and vindicated (see Radder, 1988). This "referential realism" is based on an interpretation of the history of the experimental sciences as an essentially differentiated process, in which there is an interplay between developments in the dimensions of experimental action and production, conceptual-theoretical work, and formal-mathematical activity.

As to the conceptual-theoretical side of science, a referentially realist interpretation fully accepts the conclusion of many historical studies that conceptual discontinuity is a fundamental and recurring feature of theoretical development. To be sure, not every change implies a radical conceptual break. Yet, the cumulative result of regularly occurring, partial conceptual discontinuities is an increasing conceptual divergence in the long run. This conclusion supports a form of conceptual relativism, which implies that the plausibility of the representational truth of conceptual-theoretical knowledge claims depends essentially on their sociohistorical context.

However, if we do not only look at conceptual-theoretical aspects but also take into account the formal-mathematical intertheory relations (see chapter 3) and the actual performance, or material realization, of experiments (see chapter 2), it proves to be feasible to develop a referential realism for the experimental sciences. Two basic epistemological claims concerning the reference and coreference to reality appear to be plausible:

1. We know that the descriptive terms occurring in the conceptual-theoretical description of an experiment refer to elements in a domain of a human-independent reality, if the experiment can be materially realized in a reproducible manner.[3]
2. We know that (possibly discontinuous) conceptual-theoretical terms corefer to the same element in a domain of reality, if they *both* stand in a relation of formal-mathematical correspondence *and* refer to the same, reproducible material realization.

In this summary I must confine myself to two explanatory remarks concerning the substance of this referential realism. First, because of its compatibility with conceptual relativism, this form of realism is consistently nonrepresentational: the conceptual-theoretical interpretations of science cannot be taken to "re-present" or "mirror" (either directly or approximately) an independently existing reality. A knowledge claim that meets the above criterion of reference tells us *that* the referent in question really exists; and, most importantly, when different and possibly conceptually discontinuous terms satisfy the coreference criterion, we know *that* they denote the same referent; however, these criteria do *not* reveal *what* this referent, in itself, is like in a conceptual-theoretical sense. In other words, this realism is strictly referential. As such it clearly contrasts with (and is critical of) those attempts at meeting the Kuhnian challenge that argue for a (representational) convergent realism backed up by a causal theory of reference.[4]

Second, because of the role played by conceptual-theoretical and formal-mathematical aspects, referential realism is not a form of "instrumental" or "experimental" realism.[5] Conceptual-theoretical interpretations are needed to guide and make sense of the process of materially realizing experiments; and intertheory relations of a formal-mathematical kind are a crucial requisite for a sensible definition of the notion of coreference. Yet, it will also be clear that experimentation forms an important ingredient of referential realism. In any case, in view of the discussion of the Bachelardian challenge, the present chapter focuses on the experimental aspects.

4.3 MEETING THE BACHELARDIAN CHALLENGE: AN ONTOLOGY OF PERSISTENTLY REAL POTENTIALITIES AND HISTORICALLY CONTINGENT REALIZATIONS

Let me assume, then, that referential realism constitutes an adequate answer to the Kuhnian challenge. Thus, scientific knowledge claims may refer and corefer to *elements* of a *human-independent* reality. As we have seen, experimental action and production play a central role in this form of realism, together with conceptual-theoretical and formal-mathematical work. Therefore, our meeting of the Kuhnian challenge immediately confronts us with the Bachelardian challenge: how can experimental or experimentally tested knowledge, including its theoretical explanations, "be about an independent reality," if the phenomena on which it is based (the experimental processes and results) do not occur "naturally" but have to be artificially produced and maintained in special laboratories? If not only the knowledge of experimental processes and results but even their existence is essentially dependent on the work done by human beings, how can we ever substantiate the claim that knowledge in the experimental sciences refers to a human-independent reality? In the present section, I will attempt to answer these Bachelardian questions by proposing and vindicating a philosophical-ontological specification of the previously mentioned "elements" of reality.

Now, different formulations and accounts of the relevant notions can be found in the literature. The nature of the experimental production process, for example, is variously interpreted as a matter of transcendental constitution or of human creation or of social construction or of a construction process in which also nonhuman actors play a role. Furthermore, the views about the scale of this process diverge as well. Some claim that experimental processes or results are locally produced, by particular scientists in particular laboratories, while others hold that they are essentially dependent on the work done by the sum total of all scientists and laboratories involved. However that may be, I will not deal here with all these different positions separately.[6] Instead, I will confront the Bachelardian challenge in its most general form: how is it possible for scientific knowledge to be about a human-independent reality when human work plays such an essential role in producing experimental processes and results? It will be clear that the Bachelardian challenge is not an idealist challenge, since the materials on and with which humans work may be taken into account. This alone, however, does not suffice to answer the above question. After all, in the Bachelardian view, it remains true that the products created with the help of these materials cannot be interpreted as being about a human-independent reality.

Right from the start, it should be clear that my response to the Bachelardian challenge does not imply a rejection of the claim that a lot of nontrivial work is necessary for successfully producing particular experimental processes and results. After more than a decade of intensive study of experimentation, it is evident that most concrete realizations of experimental phenomena do not exist in nature until they are artificially produced by human beings.[7] This alone, however, does not yet answer the crucial question of whether an account exclusively in terms of human work provides a sufficient and plausible interpretation of scientific truth claims.

My (negative) answer to this question is based on the distinction between a *reproduced* and a *reproducible* experiment. A reproduced experiment is a particular reproduction of an original experiment, whereas a reproducible experiment is a particular experiment that might be reproduced. In other words, a reproducible experiment points to a set of reproducibility conditions that in fact are not yet, or shall perhaps never be, realized. Consequently, a reproducible experiment possesses a nonlocal meaning, which essentially transcends the meaning of its actually realized, local reproductions. Here nonlocality stands for a specific indeterminateness by actual, local realizations and a certain openness to *particular* (but as yet unrealized) possibilities. Since nonlocal reproducibility does not imply that the experiments in question can actually be universally reproduced, it is important not to confuse nonlocality and universality.

So far I have only offered a definition of the notion of a reproducible experiment. But is it a sensible definition? I think it is. My main argument for this is that in scientific practice claims of (nonlocal) reproducibility are often made and have often met with considerable success.[8] An experimenter who claims that an experiment might be reproduced at a different location assumes (implicitly or explicitly) that the meaning of this one experiment transcends the context of its particular, local realization. Then, every successful reproduction of the experiment increases the plausibility of this assumption. Put differently, a reproducible experiment points beyond its own production context towards the possibility of reproducing it, in advance of any of its actually realized reproductions. The fact that, of course, in practice the realization of particular reproductions may fail implies that claims on reproducibility are and will always be fallible. This, however, does not detract from but rather reinforces the meaningfulness of claims on reproducibility.

In sum, my argument for claiming that "human work" is not sufficient for convincingly explaining scientific experimentation involves two steps. First, the notion of a reproducible experiment is meaningful, because the claim that an experiment is reproducible can be substantiated by successfully realizing a number of particular reproductions. Second, the meaningful distinction between reproduced and reproducible entails that it is literally impossible to *make* an experiment reproducible. In other words, reproducible experiments, in contrast to reproduced experiments, cannot be adequately explained if we restrict our accounts to the particularity of local experimental episodes.

Thus, reproducible experiments have a nonlocal meaning. The possibility of experimental reproduction, "being reproducible," is not made by the work done in any local laboratory context. This entails that ontologically we also should make a distinction between the being reproducible of an experiment and its actually realized (re)production in a particular case. While the latter is produced or constructed by particular human beings in a particular laboratory setting, the former is not.[9] Consequently, the Bachelardian challenge can be met. The above argument shows the *feasibility* of interpreting the ontological notion of "being reproducible" in a realist manner. The following is, I think, a natural and plausible development of this notion.[10]

If a (conceptual-theoretical) term from the theoretical description of a reproducible experiment refers,[11] it "is about" a *persistent potentiality of reality*, which as such is independent of the existence and knowledge of human beings. The *realization* of this potentiality, however, essentially requires human (material and theoretical) work. Successfully realizing a particular reproducible experiment depends on contingent historical conditions that just happen to occur or are made to occur. Because of this contingency, the range of these realizations cannot be established a priori but has to be investigated a posteriori by means of historical studies of the development of (experimental) science. Thus, the claim that scientific experiments can be universally reproduced, which is rather implausible on the basis of historical evidence, is not part of the views on science explained in this chapter.

Ontologically, the human-independent persistent potentialities and the contingent historical realizations are fundamentally different. Epistemologically and methodologically, however, "objective discovery" of persistent potentialities is intrinsically connected with "human production" of contingent realizations. It is this specific connection that best explains the fact that in science not anything goes and that, in particular, obtaining reproducibility usually requires a lot of (hard) work. In the above discussion, I have focussed on ontological questions. Yet, the ontological claims are not independent of epistemological arguments. As will become clearer in the next section, the ontological proposals make an essential use of a detailed analysis of experimental reproducibility. If it were the case that reproducibility did not play any epistemic role at all in scientific practice or if attempts at reproducing experiments generally failed, the proposed ontology would lose its point. However, even if ontology and epistemology are not independent, the realist interpretation proposed in this chapter excludes a reduction of ontology to epistemology. "Reality" cannot be identified with "our knowledge of reality." The potentialities of nature persist independently of our existence and of our knowledge, even if they can be realized only if we cooperate and are able and willing to do the required material and theoretical work.

To avoid misunderstanding, I want to emphasize that my claim is not that the contingent, historical realizations are in some sense unreal or illusory. Particular experimental processes and results certainly exist, just as other human products such as shoes or tables do.[12] The point of the above argumentation is, however, that their reality—in contrast to the reality of the persistent potentialities—remains essentially dependent upon the existence of human beings.[13]

Finally a matter of terminology. The locution "realization" is, I think, the most appropriate for catching the main characteristics of the mixture of discovery and production in scientific practice. Reproducible realization entails the "actualization" of some potentiality of reality, "brought about" by some human beings, who in doing so "acquire knowledge" of something by realizing—that is, understanding—it.[14] In contrast, the term "instantiation" carries with it too much of the empiricist connotation of a straightforward or "instant" apprehension of reality. "Creation," on the other hand, points too much in a constructivist direction and leaves too little room for the independence of the potentialities of nature.

4.4 REALIZING TYPES AND RANGES OF REPRODUCIBILITY

After the more general discussion of the realization of reproduced and the reality of reproducible experiments in the preceding section, I now want to enter into more detail by making use of the differentiation between types and ranges of reproducibility. In chapter 2 I distinguished two basic aspects of experimentation. The first is the (conceptual and formal-mathematical) theoretical description or interpretation of an experiment, noted as $p \Rightarrow q$. Here the description q denotes the (intended) experimental result, while the composite description p comprises all kinds of premises that are necessary for drawing the conclusion that q is the result of the overall experimental process. Second, there is the material realization of an experiment. The notion of material realization refers to the features of action and production and thus reflects the fact that science involves not only thinking, reasoning, and theorizing but also doing, manipulating, and producing. On this basis I introduced and explained a distinction between the reproducibility of the material realization, of the theoretical description, and of the theoretical result of an experiment.

The question, then, is what can be said about the intrinsic relationship between objective discovery and human production in these three different cases of reproducibility? Consider first the reproducibility of the material realization of an experiment. As we have seen in section 2.2, this type of reproducibility requires that, guided by *some* theoretical interpretation, the same actions can be performed and the same experimental situations produced from the point of view of the "common language" description of the material realization of the experiment. The crucial point is that reproductions of this type may be obtained on the basis of a number of, possibly radically different, members from a class of theoretical interpretations. What is required for materially reproducing an experiment in an individual case is, first, some theoretical description; second, the necessary experimental devices and materials; and, third, a number of manipulative skills in order to perform the experiment correctly. By definition, in this type of reproduction belief in the representational truth of the conceptual-theoretical description, though cognitively indispensable, is epistemically bracketed. Therefore, in this case the question whether the truth of this description is discovered or produced does not arise. As to the experimental devices and materials, they are definitely produced and not discovered. To make an experiment work, the devices have to be skillfully constructed and the materials carefully prepared. Moreover, due to contingent historical developments, the materials and the conditions needed for materially realizing the reproduction might simply not (or never) be available any more. The same applies to the necessary skills: acquiring and maintaining these skills may well demand a lot of special conditions, which have to be artificially created and can in no way be taken to exist independently of human existence or human knowledge. In sum, the material realization of a particular reproduction should be considered a human product.

Yet, what we do not and cannot produce is the reproducibility of this material realization. Thus, any individual, reproducible material realization of an experiment is about an independent and persistent potentiality of nature. It is this real potentiality that offers us the possibility of engaging in a learning process, in which we try to skillfully reproduce the material realization at another space-time location, perhaps on the basis of a radically different theoretical description. In epistemological terms this means that, when the material realization of an experiment has been successfully reproduced up to now, we have every reason to believe the claim that the terms from its theoretical interpretation refer to a persistent potentiality of reality, which might be realized if the appropriate conditions can be produced and maintained. This entails that referential realism—though certainly more modest than traditional forms of representational realism—allows a positive and nonlocal role for reality in the production of knowledge. In this respect it contrasts with interpretations according to which the impact of reality can be felt only negatively and locally, as a contextual "resistance" to our attempts at creating "coherence" or "convergence" within particular experimental practices.[15]

Let us now take a look at the reproducibility of an experiment under a fixed theoretical description $p \Rightarrow q$. In this case we have to deal with a repeatability of the experiment from the point of view of the theoretical interpretation in question. If a particular reproduction succeeds, the people involved believe that what has been reproduced is the experimental process as described by this theoretical description. In other words, they believe in the correctness of this description. Nevertheless, given the fact that the theoretical interpretations of experiments in the history of science regularly exhibit conceptual discontinuities, it is reasonable to consider any particular conceptual-theoretical interpretation as a construct, the plausibility of which is restricted to specific periods in the history of science. Therefore, accepting and acting upon the interpretation in question is based on specific, constructive work. Moreover, in line with the above argumentation, the specific devices, materials, and skills needed for materially realizing the experiment should also be considered as being produced through human work.

But again, what is not produced is the fact that the experiment is reproducible under the interpretation in question. In other words, the nontrivial fact that, *given* this description and these devices, materials, and skills, the experiment shows a definite and reliable insensitivity or stability with respect to translations in space and time is not constructed by us. Moreover, if the terms from the theoretical description refer (which can be checked by applying the reference criterion), they refer to a persistent potentiality of a human-independent reality.

The third type of reproducibility concerns the result of an experiment. We may, for example, have $p \Rightarrow q$ and $p' \Rightarrow q$, where p and p' are two (possibly radically) different descriptions that will mostly (but not necessarily) describe two different material procedures for obtaining q. As a matter of terminology, I speak in this case of the replicability of the experimental result. Thus, by definition, replicability entails that the same result q might be realized by means of a number of different experimental processes. Again, in a series of particular replications the separate theoretical descriptions q, p, p', etc. and the particular devices, materials, and manipulative skills required for replicating the result should be considered as produced by human beings. But, in line with the above argumentation, it is just as true that the replicability of the result transcends, and is not determined by, (the sum total of) the different individual replications. In other words, this replicability is not a human product. Consequently, being replicable cannot be adequately explained on the basis of the characteristics of the particular replication contexts. Moreover, if all the separate replications are known to satisfy the criterion of reference, the replicability of the result q implies that we have discovered an *indeterminate set of interconnected, independently persisting potentialities of reality.*

Two final remarks: First, it is important to note that two reproductions are never identical in all respects. Their similarity is relative to the way they are described. For example, in the case of a reproduction under a fixed theoretical interpretation, two reproductions may differ with respect to what is left out or deemed irrelevant by the interpretation in question. And in the case of a replication, we may even have two radically different experimental processes that replicate the same result. In the second place, what I have said in the preceding section about the range of reproducible experiments applies just as well to the different types of reproducibility. Successfully realizing a specific type of reproducibility in a particular case essentially depends on contingent historical conditions. A priori, we simply do not know how far these conditions will, or can be made to, range in space and time.

4.5 THE ABSTRACTION (PLUS INTERPRETATION AND REALIZATION) OF NONLOCALS

It is interesting to analyze the notion of replicability in more detail, since it exhibits some significant differences with the other two types of reproducibility. These differences come to the fore when we ask what it means to call an experimental result replicable. Suppose it is claimed that the result q, so far obtained by means of two different experimental processes $p \Rightarrow q$ and $p' \Rightarrow q$, is replicable. This claim means that q might be reproduced in at least one more experimental context that may be radically different from p and p'. For this reason it is not adequate to interpret the inference of the replicability of q as a matter of induction. Doing so would obscure the differences in the productive work required for obtaining the replications.[16] In this respect there is a discrepancy between replicability on the one hand and reproducibility of the material realization or reproducibility under a fixed interpretation on the other. In the latter cases the claim that the experiments are (nonlocally) reproducible does at least seem to be analogous to inductive claims.

In contrast, claims on replicability are not a matter of induction (and neither of abduction nor idealization) but rather of *abstraction*. As argued above, reproducible experiments have a nonlocal meaning. More particularly, when a result q is replicable in different experiments, this means that q may obtain independently of the specific experimental processes in which it has been realized so far.[17] The replicability of q is not constituted by (the sum total of) these processes. Therefore, in claiming the replicability of a result, we abstract from these particular experimental processes. That such claims are frequently made and confirmed in experimental practice is an undeniable fact. Thus, at least we have to face a possible explanation of this fact.

The explanation advocated here can be linked to the old question of the status of universals. As argued above, on the basis of a study of scientific practice we can only claim that replicable realizations of experiments have a nonlocal, and not a universal, range. Let me therefore rephrase the above question into that of the status of "nonlocals." In this case, nonlocals are experimental results, the meaning of which essentially transcends their local production contexts. The account of experimental replicability described in the preceding section amounts to a realist interpretation of nonlocals. If the terms from a particular theoretical description refer to a real and persistent potentiality of nature, then, of course, the same is true of the result q. Furthermore, when it is also the case that the abstracted result q is replicable, then q refers to an indeterminate set of interconnected potentialities (which might become realized in different individual cases, if the right material and theoretical conditions can be produced and maintained). Thus, this account bears some analogies to the Aristotelian position in the debate on universals (*universalia in rebus*), in the sense that the nonlocal replicability of experimental results is accounted for on the basis of the independently persisting potentialities that are inherent in the things themselves.[18]

Of course, in every individual replication we not only abstract the result but also theoretically interpret and materially realize the overall experimental process in question. All three activities—interpretation, material realization, and abstraction—play a vital but different role in scientific practice. Since the theoretical interpretation guides and makes sense of the material realization, experimentally generated experience is not "given" but theory-laden from the start. Yet, the identity of a material realization is not constructed or constituted by any of its particular theoretical interpretations. The material realization of an experiment has a relative autonomy, without which theoretical controversy concerning this experiment would be unintelligible and even impossible. Finally, abstraction through replication involves disregarding the (always existing and often sizeable) variances between individual replications. In replications we do not just disregard the particular space-time locations (as we already do in reproducing an experiment under a fixed theoretical description), but we also abstract from (a large part of) the particular experimental processes by which the results in question have been produced so far. Thus, replication exemplifies the nonlocal aspect of scientific experimentation most clearly. Abstraction through replication enables us to systematically conceptualize experimental results arising in essentially different situations. As such it constitutes an important step towards theory formation.

In recent philosophy, however, the notion of abstraction has become definitely unpopular. In the history of philosophy the abstract has mostly been contrasted with the (presumably) directly given, concrete particular. Abstraction would be some kind of inference from uninterpreted, particular experiences to (psychologically, linguistically, or theoretically) interpreted general concepts (cf. Blokhuis, 1985). A classical adherent of such views is Locke. He defines abstraction as a process in which "ideas taken from particular beings become general representatives of all of the same kind."[19] Since Kant's Copernican turn, however, many philosophers have argued that concept formation does not start from uninterpreted experience. On the contrary, they say, it is the hypothetically applied universal concepts that constitute an individual experience as an experience *of something*. In "abstracting" from the particularities of individual cases, it is these concepts that tell us what counts as similar or dissimilar, and what as relevant or irrelevant.[20] As a consequence, in contemporary philosophy the focus is on the constitutive activities (by the subject, by the linguistic and scientific community, or by the relevant actors), whereas the notion of abstraction has virtually disappeared from philosophical discourse.

The views on science developed in this chapter require a reconsideration of this issue. What is at stake is, in my terms, the relationship between the particular and the nonlocal. More specifically, is the particular constituted by the nonlocal, or is the nonlocal abstracted from the particular? My answer to this question has to be differentiated. On the one hand, I fully endorse the modern idea that the experimental replication of a particular result q presupposes some conceptual-theoretical interpretation of the form $p \Rightarrow q$. On the other hand, the nonlocal replicability of this (theoretically interpreted) result essentially transcends the sum total of all the specific replications carried out so far. Therefore, the replicable nonlocal q is obtained by abstracting from the set of particular replications. The latter fact has been unjustly lost in many modern interpretations of science.[21]

Having said this, it is equally crucial not to relapse into a view of science in which the meaning of replicable results is fully severed from experimental practice. After all, distinguishing replication and replicability is not the same as disconnecting them. Moreover, the notion of a potentiality would be vacuous if it were not based on successfully performed, concrete realizations. In other words, to abstract in the sense intended here should not be interpreted as to separate (the abstract from the concrete, or the nonlocal from the local). Instead, abstraction through replication should be seen as an activity that, in being rooted in a fundamental indeterminacy that is inherent in the results of experimental processes, reflects a sensitivity to as yet unrealized possibilities.

4.6 BETWEEN TRANSCENDENTAL REALISM AND CONSTRUCTIVISM

Since the main aim of this chapter is to present a new, plausible response to the Bachelardian challenge, I have so far not said much about comparable or alternative views. But of course, the account in terms of an ontological differentiation between independently persisting, real potentialities and historically situated, particular realizations has not been developed in a vacuum. The idea of ontological differentiation has a long history, going back as far as the philosophy of Aristotle. In our century, various authors, including Whitehead, Popper, Heisenberg, Harré, Bhaskar, and Cartwright, have proposed ontologies in which the real is not exhausted by the actual. Depending on their philosophical positions and on the nature of the philosophical problems to which they responded, a large number of different terms characterizing the domain of the nonactual have been put forward: potentialities, dispositions, propensities, powers, liabilities, tendencies, affordances, and capacities. The philosophical significance of the productive character of experimentation has been pointed out as well by a number of authors. Apart from Bachelard, I should mention here Habermas, Bhaskar, Collins, Latour, Woolgar, and Hacking. Again, sundry notions have been proposed. One speaks variously of the experimental constitution of objectivity, of the production of closed systems, of the (social) construction of scientific facts, or of the creation of experimental phenomena. This is not the appropriate place to discuss all these different positions and notions. I want to make two exceptions, though. In order to further clarify and properly situate my own views, I will briefly consider those of Bhaskar and of Woolgar and Latour.

In my view it was Bhaskar who formulated the Bachelardian challenge most pointedly. He moreover responded to it explicitly by developing a particular blend of ontological or transcendental realism and epistemological relativism (see Bhaskar, 1975, 1978). Although his work is thoughtful and stimulating, it also suffers from a number of rather fundamental problems (cf. also Radder, 1988, 77-79). Here I will deal briefly with the following three issues.

First, Bhaskar speaks of a stratified reality, in which specific and enduring tendencies are operative, independently of their regular manifestations in the closed experimental systems produced by human beings. He moreover claims that it is these real tendencies that "generate the phenomena," and that "the order discovered in nature exists independently of men, i.e. of human activity in general." (Bhaskar, 1978, 25 and 27). This at least strongly suggests an ontological hierarchy, in which the enduring tendencies, or natural laws, are primary, since they embody an independent order and are the originators of the phenomena. In contrast, the notion of human-independent potentialities is less demanding. These potentialities do not generate but enable the realization of the experimental phenomena. Furthermore, the order in nature is not simply discovered, but results from a process in which objective discovery and human production are intrinsically connected. For these reasons, there is no ontological stratification or hierarchy between the persistent potentialities of reality and their contingent, historical realizations. Such a view agrees with Keller's critique of the dominant, hierarchical interpretation of the laws of nature.[22] And it is concordant with her account of dynamic objectivity as aiming at "a form of knowledge that grants to the world around us its independent integrity but does so in a way that remains cognizant of, indeed relies on, our connectivity with that world" (Keller, 1985, 117).

Second, I have pointed out that, contrary to Bhaskar's claim, reproducible experiments, or experimental regularities, are *not* produced by scientists. Therefore, I do not make use of Bhaskar's rigid and questionable ontological distinction between closed and open systems, which plays a central role in his philosophy (Bhaskar, 1978, esp. ch. 2). One particularly implausible implication of this distinction is that reliable prediction is exclusively possible in closed (laboratory) systems and not at all in open (technological) systems. Thus, Bhaskar's view makes any prospective technology policy at one stroke nonsensical. In my view open, in the sense of completely unpredictable, systems do not exist, since no system is radically unstable or totally chaotic in *all* respects. By implication, also closed systems should be conceptualized differently.[23]

Third, the analyses in the preceding sections imply that making good knowledge claims in experimental science necessitates the realization of the reproducibility conditions pertaining to the material realization and the theoretical interpretation of the experiments in question. This means that knowledge and power, although not identical, are intrinsically connected. First, realizing knowledge *requires* (productive) power; and second, knowledge *is* (repressive) power when its realization, as is often the case due to conditions of scarcity or intolerance, excludes desirable, alternative realizations (see chapter 6; cf. also Latour, 1983, 1987b; Rouse, 1987). At this point a further difference with Bhaskar's views shows up. He writes that

> the philosophical position developed in this study does not depend on an arbitrary definition of science, but rather upon the intelligibility of certain universally recognized, if inadequately analyzed, scientific activities. (Bhaskar, 1978, 24)

In his interpretation of science, Bhaskar assumes that experimentation, being such an activity, is universally acknowledged as a legitimate method for producing reliable knowledge. This, however, is not the case.[24] Experimenting is not *the* natural procedure for obtaining justified knowledge. The existence and acceptance of the experimental method as a legitimate way of producing knowledge is no more and no less than a nonlocal pattern in the historical development of most modern sciences. As such, it offers the opportunity of realizing (and exploiting) certain potentialities of reality, *if and only if* the people involved are willing or can be made to exercise and endure the required power.

A further consequence of the approach taken in this chapter is that so far I have not been bothered too much about skeptical reinforcements of Bachelard's claims. And in so far as a coherently developed skepticism cannot be refuted by logical means, this is all the better. Yet, the impossibility of compelling refutation does not mean that arguments for or against the plausibility of philosophical interpretations of science are senseless. In the remainder of this section, I will illustrate this by briefly discussing the constructivist development of Bachelard's views by Latour and Woolgar. Consider first the following claims:

> the spectrum produced by a nuclear magnetic resonance . . . spectrometer . . . would not exist but for the spectrometer. It is not simply that phenomena *depend on* certain material instrumentation; rather, the phenomena *are thoroughly constituted by* the material setting of the laboratory. The artificial reality, which participants describe in terms of an objective entity, has in fact been constructed by the use of inscription devices. Such a reality, which Bachelard . . . terms the "phenomenotechnique," takes on the appearance of a phenomenon by virtue of its construction through material techniques. (Latour and Woolgar, 1979, 64)

Let me call this position "strict localism." It claims that it is the particular, local laboratory conditions that fully account for the existence and meaning of the relevant phenomena.

Now, even if this position is not logically incoherent, it seems to me to be quite implausible. A crucial problem is that it implies the impossibility of genuine reproductions, in the sense that, in terms of the relevant description, re-productions create *the same* processes or results in a different context. After all, if a reproduction were a re-production, this would impose an outside constraint on the reproduction context, in the sense that the same process or result should be produced. In that case the reproduced phenomenon would possess a broader meaning that would not be fully constituted by the local material setting of the reproduction context. Strict localism, in contrast, implies that what counts as a reproduction in a particular local context will be completely determined by the characteristics of that context itself. This implication, however, makes it a complete mystery how and why scientists from different local contexts come to agree on issues of reproducibility, as they frequently do.[25] More generally, a strictly localist point of view faces fundamental problems in accounting for all broader patterns in scientific developments.[26]

In his more recent work, Latour has diagnosed this problem and moved on to a different form of constructivism that seems to distance itself somewhat from strict localism. Now, testing scientific predictions, including predicted reproductions, is interpreted as a process of "translation" or "extension," from one laboratory to another or from a laboratory to a technological setting. Latour's new explanation of the success of reproductions in different contexts is that, in fact, the contexts are *not* different because they have been *made* the same in an (apparently successful) process of translation. Since the production and the reproduction contexts are the same, a successfully *pre*dicted reproduction is, according to Latour (1987b, 249), in fact a *retro*diction. The difference with a strictly localist point of view is clear: making the two contexts the same cannot be achieved if one stays in and sticks to one of them, because it requires a literal extension from the production to the reproduction context in actual space and time.

This new approach to the issue of nonlocal reproducibility, however, engenders new problems, which are as difficult to solve as the old ones. As I have noted above and as is proven by the existence of different types of reproducibility, the original and the reproduction context always and necessarily differ in more or less respects. Although in principle the point can be made with reference to all three types of reproduction, it is the practice of replicating a result q by means of two radically different experimental processes p and p' that provides the stiffest counterexample against the claimed identification of prediction and retrodiction. Since all knowledge production is a matter of transformation, the replicating process $p' \Rightarrow q$ will in general be an extension of *some* earlier experiments, even if this specific overall process is realized here for the first time. In this sense Latour is right. But the critical point is that the overall replicating process is *not* an extension of, but radically different from, the overall replicated process. Clearly, Latour's explanation of the success of predicted reproductions, and more generally his account of how science manages to "act at a distance," does not work. Consequently, it does not constitute a viable alternative to the ontology of independently real potentialities and their historically contingent realizations.

4.7 EXPERIMENTATION VERSUS OBSERVATION?

A final question to consider concerns the *scope* of the epistemological and ontological proposals made in the preceding sections. So far, my claims have been restricted to experiment and experimental sciences. But what about (pure) observation and observational sciences? Is it appropriate to speak of the observational realization of human-independent potentialities, to which the terms from the theoretical description of materially reproducible observations refer? In order to answer this question, I will first define scientific observation by distinguishing it from both experimentation and perception.

As we have seen, experimenting involves material realization and (conceptual and formal-mathematical) theoretical interpretation. In general, the same is true of scientific observation. Consider, for instance, the observation of the oblateness of the sun. Pinch presents the following account of one such observation dating from 1967:

> Dicke and two Princeton colleagues, Mark Goldenberg and Henry Hill, designed and built a telescope capable of measuring the solar oblateness to a sufficient accuracy to test relativity theory. . . . The telescope consists of a series of mirrors and lenses which track the Sun and provide a solar image. This image is projected on to an occulting disc. A small portion projects beyond the disc and is passed through a rapidly rotating scanning wheel perforated by two diametrically opposed apertures. The light passing through these apertures is detected by a photoelectric cell. If the Sun is oblate, slightly more light will be transmitted to the photocell when the aperture in the wheel samples light from equatorial regions. The radial difference . . . is, in effect, measured 244 times per second.[27]

It will be obvious that an observation such as this one, which is typical of modern science, involves both an extensive material realization and a lot of theoretical interpretation. The only difference with a "genuine experiment" is that the observed (feature of the) object in question, the sun and its oblateness, is not itself materially realized by human interference. This I will take as the characteristic distinction between observations and experiments: in observation processes, the observed objects are not (directly) materially realized by human interference.[28] Moreover, in a number of cases, the observed objects will, in principle or in practice, be completely outside the scope of human manipulability. That is to say, they are not materially realizable at all. This holds true for many astronomical and astrophysical observations, as well as for numerous observations in the earth and biomedical sciences. Here I will restrict myself to such "not materially realizable observed objects," since they constitute the relevant cases for the philosophical questions under discussion.

Next, it is just as important to distinguish observation from sense perception. Perception surely plays a role in observational processes. More particularly, at some stages of the observational process an interaction with our sense organs is needed so that we can become aware of certain outcomes of the process. But, as is clearly illustrated by the example of the observation of solar oblateness, realizing the overall observational process involves *much more* than just perceiving these outcomes. Pinch (1985, 8) rightly concludes:

It seems that the process of observation has become refined in such a way that what we might call "primitive sense perception," or the output of an "inscription device," forms only the last part of a chain of inference.... the process of observation in modern science is one in which experimental practices and theoretical interpretations take on central importance.

Consequently, observation, and a fortiori experimentation, cannot be reduced to sense perception. The central notion of empiricist philosophy of science and epistemology is doubly inadequate: first, because it fails to capture the significance of theoretical interpretation; and second, because it neglects the role of active and skillful intervention in the practice of materially realizing both experiments and observations.

After these terminological clarifications, let us confront the central question of this section: does the fact that an observed object is not itself materially realizable make an epistemological difference? For Hacking it does. Since such an observed entity cannot be manipulated (as is the case in astrophysical observation), we cannot have the proper reasons for believing in its reality:

> Long-lived theoretical entities, which don't end up being manipulated, commonly turn out to have been wonderful mistakes. (Hacking, 1983b, 275)

And, concerning astrophysics in particular:

> People, it has been said, are tool-making animals. When we use entities as tools, as instruments of inquiry, we are entitled to regard them as real. But we cannot do that with the objects of astrophysics. Astrophysics is almost the only human domain where we have profound, intricate knowledge, and in which we can be no more than what van Fraassen calls constructive empiricists. (Hacking, 1989a, 578)

I think, however, that drawing such a sharp dividing line between experimenting and observing is not plausible, for the following reasons. For a start, not only black holes and gravitational lenses (Hacking's examples), but also stars and even "our" star, the sun, cannot be manipulated in principle. Are we not entitled to regard the sun as a real entity?

Furthermore, the reason why Hacking restricts his realist interpretation to manipulable entities is that he assumes a fundamental distinction between manipulating, or "experimenting with," entities and "experimenting on" entities.[29] More specifically, his assumption is that "experimenting with" merely requires a few "home truths" about the entities in question. These home truths would remain stable across time, because they are independent of the ever changing high theories and theoretical models that are used in the description of the hypothetical entities *on* which we experiment. Indeed, it is precisely this claimed stability that forms a central element of Hacking's response to the Kuhnian challenge. However, both the claim of the independence of "experimenting with" from high theory and the claim of the stability of home truths have been convincingly criticized.[30] Consequently, this criticism has at the same time undermined Hacking's main reason for making the contrast between experimentation and observation in the first place.

Finally, Hacking's argument seems to employ an overly literal use of the notions of manipulation and tool. After all, not all entities that we may standardly use as tools for inquiry are, literally, manipulated or manipulable. For example, in an experiment, we may use electric power generated by a standardized solar or wind energy device. In such a case the sun or the wind are not being "manipulated" in Hacking's sense of the word. Yet, they function as essential parts of the technological tool. Moreover, in this particular case, the "observational" tool may well be more stable and reliable than an "experimental" alternative, such as a nuclear energy reactor.

My conclusion is that Hacking's arguments are not sufficient to sustain a fundamental and epistemologically decisive contrast between observational and experimental knowledge claims. Therefore, I suggest that the epistemological and ontological views put forward in the preceding sections may be plausibly extended to include knowledge claims produced by or based on observation. That is to say, it is reasonable to believe that the terms from the theoretical description of an observation that can be reproducibly materially realized refer to elements of an independently existing reality, just as the terms from the description of a materially reproducible experiment do. Moreover, analogous to the ontology for experimental science, we may claim that the "elements" of this observed reality can be specified as being independently persisting potentialities, which might be realized if the appropriate observational conditions can be produced and maintained.

CHAPTER 5

Normative Reflexions on Constructivist Approaches to Science and Technology

5.1 INTRODUCTION

As I have hinted at before and as I will explain in more detail in chapter 8, philosophy is (or should be) at once theoretical, normative, and reflexive. So far, the focus has been on theoretical-philosophical explanations and interpretations of experimental and theoretical science. In this chapter, I will deal with certain normative and reflexive philosophical issues. More in particular, I will offer a number of normative reflexions on constructivist approaches to science and technology. As we will see, several of the theoretical issues discussed earlier—such as nonlocality, realization, and reality—will again play a significant role in the argumentations set out here.

Normative questions deal with what ought to be done or believed. By *normative reflexion* on certain activities or ideas I mean: thinking explicitly and fundamentally about the normative or normatively relevant presuppositions, assumptions, and implications of those activities or ideas. In the case of the study of science and technology, normative reflexion means developing systematic analyses and evaluations of its relevance for normative questions in our technoscientific world. Normative reflexion is significant because science and technology studies do not simply mirror this world. Like all other intellectual activities, they are themselves necessarily part of it: they shape it and are shaped by it.

Since the beginning of the systematic social study of science and technology, normative reflexion has often been an integral part of it. We may think of the Bernal-Polanyi debate and the "critical theory" of the 1930s; the views of the student movement in the 1960s; or the ideas of Marxist or radical scientists and those of environmental or feminist critics of science and technology in the 1970s. Since about 1975, the social study of science and technology has become ever more established. At the same time, the role of explicit normative reflexion has steadily been decreasing.

In constructivist approaches, to which I will limit myself in this chapter, normative reflexivity is nearly absent.[1] But we should also note that, for instance, policy or technology assessment studies are not, by themselves, normatively reflexive in the sense defined above. Although these studies are, in part, concerned with normative issues, they do not necessarily or even often reflect in an explicit and systematic way on their *own* normative role. In other words, a normatively relevant study of science and technology does not automatically involve normative reflexion.

Below I will provide a number of normative reflexions on the highly influential constructivist approaches to science and technology. In particular, I will deal with the cognitive, social, or ethical norms and values embodied in both instrumental and reflexive ethnography, in social constructivist programs for the study of science and technology and in the so-called actor-network theory of technoscience.[2] In doing so, I will examine questions like the following: Which methods are recommended (or criticized) with respect to the study of science and technology, and what are the normative implications of these recommendations or criticisms? Which types of solutions to social problems of our technoscientific world are possible or impossible, given certain constructivist views of science and technology? Which actors are regarded as being, or not being, involved in technoscientific developments; and whose concerns should, or should not, therefore be taken into account? As we will see, the different approaches within the constructivist camp do not always imply the same answers to questions like these. This is why my treatment of the issues in question will also be differentiated.

The first aim of the current chapter, then, is to offer a number of *analytical* normative reflexions—that is, systematic analyses of the normative or normatively relevant presuppositions, assumptions, and implications of constructivist views on science and technology. In the second place, I will argue that constructivism, in a number of cases, has consequences that are normatively questionable. In other words, I will also engage in what may be called *critical* normative reflexion. At the same time, I will propose certain alternatives to constructivist doctrines and methods, which seem more adequate from a normative point of view. In other words, the normative reflexions offered below are not merely critical but also *constructive*. Concerning the latter type of reflexion, my intention is certainly not to return to earlier views, such as the many abstract normative proposals of philosophers of science and technology or the various grandiose theories of history and society of social theorists. I think one should rather try to combine the achievements of the empirical approach—their more adequate views on the practice of science and technology—with normative insights concerning the problematic aspects of our technoscientific world.

5.2 NORMATIVITY IN CONSTRUCTIVISM

The argument of this chapter is certainly not that normative or normatively relevant views are absent from present-day constructivism. In fact, such views can be encountered in several forms. *First*, explicitly evaluative and normative claims are occasionally made. For instance, despite the fact that Collins pictures the closure of scientific controversies as being completely dependent on idiosyncratic negotiation processes, he assures us that:

> Creative science is a marvellous and beautiful exercise of intellect, ingenuity and skill. It would be absurdly arrogant for any sociologist to criticise its internal operation and thus claim to know how to do science better than the practitioners themselves. (Collins, 1975, 224)

Thus, Collins asks sociologists of scientific knowledge to respect the norm of not criticizing science. Barnes and Bloor, to take another example, argue that a relativist approach is required for a truly scientific understanding of forms of knowledge:

> If relativism has any appeal at all, it will be to those who wish to engage in that eccentric activity called "disinterested research." (Barnes and Bloor, 1982, 47)

Remarkable though this statement from the founding fathers of the "interest approach" may be, it clearly involves a normative recommendation: if we are committed to disinterested research, as everybody should be according to Barnes and Bloor, a relativist view is the obvious one. A final illustration is provided by Latour, who appends the following normative call to his Machiavellian analysis of the seamless web of science, technology, and society:

> If science and techniques are politics pursued with other means, then the only way to pursue democracy is to get inside science and techniques.... This is where we should stand if the Prince is to be more than a few individuals, if it is to be called "the People." (Latour, 1987a, 30-31)

Explicit though they are, these normative points of view are not normative reflexions in the sense explained above. They are not systematically developed and discussed but merely briefly stated in footnotes, introductions, and conclusions.

Second, a number of studies in this area may be viewed as directly relevant to normative issues because of the topics they deal with. Thus, certain studies of domestic technology (for example, Schwartz Cowan, 1985) are relevant to, and are in fact employed in, feminist critiques of dominant conceptions of gender-specific roles in western societies. Research into military technology, to mention another example, may be used to trace opportunities for disarmament, as is hinted at by MacKenzie (1987, 218). Moreover, it is sometimes noted that an undeniably important area, such as military science (see MacKenzie, 1986), or a clearly significant subject, such as the role of gender in knowledge production (see Keller, 1988), is regularly underresearched or neglected in constructivist approaches to science. Studies like the ones cited above might lead to normative reflexions, if the issues in question were to be taken up in a systematic manner.

Third, constructivist approaches may imply normative critique as a result of their theoretical assumptions. Consider social constructivism, for instance, in the form of the well-known "Empirical Programme Of Relativism" (Collins, 1981, 1983), pertaining to science, and the "Social Construction Of Technology" program (Pinch and Bijker, 1984). Social constructivism, like any theoretical position, is not normatively neutral, but its main critique is deconstructive. Philosophical views claiming that, generally speaking, acceptance of scientific statements or technological artifacts is based on their objective truth or their objective working are criticized as being false. On the positive side, the "technical" content of science and technology is deconstructed by showing that it is completely dependent on social negotiations and interests.

Starting from assumptions compatible with or inspired by social constructivism, some feminist critics, for example, have argued that notions and theories of masculinity and femininity in the biomedical sciences are not so much logical consequences of experimental findings but rather social constructions, the plausibility of which derives largely from continuing, undesirable cultural biases.[3] It has also been claimed that social constructivism has normatively relevant implications for technology (Pinch and Bijker, 1986, esp. 354–357). Since the functioning of a technological artifact is completely determined by its social context, social constructivism would clear the way for a legitimate consideration of alternatives to the presently dominant forms of technology. In this sense, this position can be normatively relevant to social choices concerning technology (cf., e.g., Noble, 1979).

I agree that these critical normative applications of social constructivism are legitimate and valuable in a number of cases. Nevertheless, problems arise when we try to move on towards a more constructive reflexivity. As I will argue below, these problems can be solved by mitigating the relativist and social reductionist philosophy of social constructivism.

Fourth, certain "cognitive" norms are proposed as being especially or exclusively appropriate to science and technology studies. The constructivist sociology of scientific facts and technological artifacts strongly emphasizes case studies and empirical work. At the same time, it shows a distrust or even an outright rejection of theoretical analyses and philosophical assessments. The presence of methodological debates within this "new empiricism" implies that the claim made above should be slightly qualified: there is some reflexion among constructivist students of science and technology on what constitutes the appropriate methodology. However, a major problem of these reflexions is that they try to maintain a sharp separation between the cognitive and the wider (social, ethical, aesthetic) normative implications of their methodological prescriptions. Consider, for example, the following claims by Callon, Law, and Rip (1986, 15):

> The act of entering the laboratory and coolly describing what one sees does not imply that sides have to be taken for or against science and technology, even if such descriptions tend to be corrosive of established myths. As Machiavelli's work shows, description serves to enhance the efficacy of *all* parties: scientists, technologists, policy-makers, pressure groups; all may profit from it.

With regard to the interpretation of their own research activities, these authors are remarkably close to a *very* standard view of science: they simply report what they "see," whereas values come into play only afterwards, in the "application" of the results by different parties. In addition, when explaining their methodology, Callon, Law, and Rip seem to suggest that normative reflexion amounts to a simple all-or-nothing matter:

> As we follow the actors of science there is no taking sides, no question of charging science with all the sins of the world or pleading its complete innocence. This is not our intention and neither should it be. (Callon, Law, and Rip, 1986, 6)

Given these views, which are certainly not untypical of recent constructivism, one of the goals of the present chapter is to show that the "cognitive" norms advocated by constructivism do have important wider implications and that, in a number of cases, there are good grounds for questioning these implications. At the same time, this will demonstrate that normative reflexion is not a matter of either wholeheartedly embracing or completely rejecting science or technology but that a much more differentiated approach is possible and meaningful. In doing so I will deal primarily with the theoretical and methodological assumptions and conclusions of constructivists; accordingly, I shall give less attention to the often interesting empirical work.

5.3 REFLEXIVITY IN CONSTRUCTIVISM

So far I have argued that, unavoidably, normative views are expressed in constructivism but that they do not amount to full-fledged normative *reflexions*. Similarly, it is true that, albeit only recently, attention has been paid to the problem of reflexivity in constructivist studies. However, the way in which this problem has been handled is far from satisfactory and clearly falls short of the requirements for systematic *normative* reflexions. In dealing with the issues in question, I will focus on the views of Woolgar, as one of the leading advocates of the recent reflexive program.

Woolgar starts from the premise that constructivist studies of science have shown that the objects do not determine their representations but that it is the other way round: the representations determine the objects. More precisely, the representations *are* the objects, and representation is all there is. Now, for Woolgar, "being reflexive" means that the constructivists' criticism of the dualism between "representation" and "object" should not only be directed at the claims of scientists but also at constructivist sociology itself. From his point of view (that this dualism is an ideology and that in fact there are nothing but representations), he faces the problem of the status of the representations of constructivist sociological knowledge. Because this fundamental problem has not yet been systematically addressed, Woolgar proposes a next step to be taken in the sociology of scientific knowledge, in the form of developing a "reflexive ethnography," the subject of which is the practice of representation itself.

> In order to come to terms with the way in which representation pervades science, our approach should be reflexive since we need to explore ways of investigating our own use of representation. (Woolgar, 1988a, 92)

This reflexivity should be "constitutive"—that is, it should apply to both content and methods of the sociology of scientific knowledge. Not only substantive constructivist claims but also the representational research methods of constructivists should be systematically questioned. Some techniques for "interrogating" the practice of sociological representation are demonstrated in the volume *Knowledge and Reflexivity* (Woolgar, 1988b). For instance, each article is followed by an editorial reflexion that reveals and highlights the rhetorical resources deployed and proposes alternative interpretations. The intended aim of these techniques is to make visible and hence neutralize the constructive, representational achievements of the authors.

On the one hand, Woolgar points out that representation, being part and parcel of the human condition, cannot be avoided. People are, and will always be, engaged in making re-presentations—that is, in distinguishing objects and their representations. On the other hand, what can and should be avoided is the *ideology* of representation, the set of beliefs and practices that are rooted in the dualistic notion that the objects preexist and determine the representations. It is this ideology that, according to Woolgar, forms the basis of the cultural and political hegemony of science. The task of a reflexive ethnography is "not just to understand the moral order which sustains the ideology of representation, but also to seek ways of changing it" (Woolgar, 1988a, 105). Woolgar suggests that only by systematically deconstructing the practices of representation can we provide a fundamental critique of the hegemony, the perceived superiority of scientific discourse (1988a, 94 and 34).

Problems of Constitutive Reflexivity

Since I also argue for a more reflexive study of science and technology, I agree with Woolgar's project in terms of the need for reflexivity. Yet, as I will show now, the claim that this reflexivity should be focussed exclusively on the practice of representation is questionable and anyhow much too restrictive.

First, the dualism between representation and object, which is the objectionable aspect for Woolgar, is not exclusively characteristic of scientific discourse. It is, as Woolgar (1988a, 101 and 103–105) realizes, also a pervasive feature of the interpretative practices of everyday life.[4] Therefore, it remains unclear why deconstructing representation "as such" is *the* appropriate strategy for a criticism of the currently dominant position of scientific discourse. Put differently, if the dualism between representation and object is a central feature of both scientific and daily life interpretations, how can it be accountable for the hegemony of the former over the latter?

Next, there is the question of why a critique and a change of the ideology of representation is considered so important by Woolgar. After all, from his point of view the claim that "representations are about a pre-existing world" is simply one more element of a discourse. This claim cannot be "wrong" or "illusory," in the sense of being untrue of reality. For this reason the question of what *is* wrong with claims like this remains to be answered by Woolgar. In a recent paper, he tries to answer this question by stating that the major achievement of reflexive deconstruction is that it imparts a continuing "argumentative dynamic" to the social study of science and technology (Woolgar, 1991). Yet he is neither able nor willing to claim that the successive steps within this dynamic represent definite advances (see Woolgar and Ashmore, 1988, 10). Apparently, "being dynamical" is recommended as an important end in itself. To me all this sounds too much like a yuppy commercial: "be dynamic and join the exciting world of constitutive reflexivity!" On this point, there is an interesting parallel with Rorty's (1979) position. Both Woolgar and Rorty assure us in an abstract way that we will be better off by giving up the idea of representing or mirroring. But as long as they do not make clear how their approach may make a difference with respect to concrete normative issues concerning science and technology, it is impossible to know what may be gained by following up their advice.

Third, it seems to me that the project of constitutive reflexivity is still caught by the same dichotomy as classical epistemology was in its search for certainty. After all, both projects start from the premise that a claim is either completely justified or inherently arbitrary. But whereas the classical epistemologist opts for an absolutistic justificationism, Woolgar endorses a judgmental relativism according to which any claim is as plausible or as implausible as any other claim. In sociological terms, his position amounts to a radical voluntarism. This is not the place to systematically criticize these epistemological and sociological theses. Instead I will restrict myself to noting their consequences with respect to normative issues. Since both analytical and critical or constructive reflexion requires making positive claims, in one way or another, consistent constitutive reflexivity is incompatible with normative reflexion.

Consider, for instance, the following two replies to the question of why deconstructing representation is important. The first runs like this: Even if we necessarily have to make use of representational practices, experience teaches us that in fact reality is too complex to be caught within one type of representation. Therefore, sticking to one type of representation implies dogmatically restricting the richness and creativity of reality. For this reason, a reflexive critique of representations is required.[5] A second answer is proposed by Mol. In a review of *Knowledge and Reflexivity*, she suggests that constitutive reflexion may contribute to redressing the balance between scientific experts and lay people, this relation being "one of the crucial political relations of the twentieth century" (Mol, 1989, 259). Unfortunately, for a consistent constitutive reflexionist such normatively relevant answers are barred. After all, the positive claims that "reality is complex" or that "there is a crucial political asymmetry between experts and lay people in the twentieth century" are simply two more opportunities for deconstruction and not a motive for normatively relevant reflexion.

Of course, since both judgmental relativism and radical sociological voluntarism are impossible in practice, it is not feasible to *be* a consistent constitutive reflexionist. For example, a basic premise of Woolgar's program is the claim that "the social network constitutes the object" (Woolgar, 1988a, 65). Questioning *this* claim would make life for the constitutive reflexionist very hard indeed. After all, as Wittgenstein remarked long ago:

> Doubting and non-doubting behaviour: There is the first only if there is the second. . . . My *life* consists in my being content to accept many things. (Wittgenstein, 1974, 46e and 44e)

Nevertheless, fostering and acting upon the illusion of being a consistent constitutive reflexionist makes one blind to the normative or normatively relevant presuppositions, assumptions, and implications of one's own position and prevents one from engaging in critical and constructive normative reflexion.

As a simple illustration, consider the following argument concerning the role of power relations in the writing practice of constitutive reflexionists. In the present situation, a claim in an English language journal gains more credibility, generally speaking, than exactly the same claim in a French or Dutch journal. Apparently, although all discourses are equal, some are more equal than others. Given this situation, an important question is: what would be the consequences of a general acceptance of the program of constitutive reflexivity? It is well known that this program demands high stylistic and rhetorical writing skills. Obviously, the vast majority of nonnative speakers will not be able to meet these demands in their English language papers, while at the same time they are under pressure to publish in English.[6] Is this potential implication of constitutive reflexivity also a desirable one? Whatever the answer, a genuinely reflexive sociology of science and technology should systematically deal with issues like this.

A final problem of Woolgar's project is that focussing reflexivity exclusively on representation is much too restrictive. When separated from its social reductionist, antirealist metaphysics, constitutive reflexivity may be valuable as a strategy for criticizing *particular* representations. But it certainly cannot be the one and only strategy for normative reflexions. The agenda of a normatively relevant, reflexive study of science and technology should include much more than merely a critique of the ideology of representation. In order to demonstrate this, I will, in the next three sections, present a number of detailed normative reflexions on specific topics within constructivist approaches to science and technology.

5.4 LOCALITY

I will start with the issue of locality. The ethnographic tradition, in particular, has stressed the local character of the construction of scientific and technological facts and artifacts (cf. Latour and Woolgar, 1979; Knorr-Cetina, 1983; Lynch, Livingston, and Garfinkel, 1983; Woolgar, 1988a). The validity of scientific knowledge and the working of technological artifacts depends completely, so it is claimed, on the idiosyncratic features of the local situation in which the knowledge and the artifacts are produced or used. More specifically, nonlocal rules do not play any guiding or constraining role in the construction of facts and artifacts, since they merely result from a retrospective rationalization of already established practice.

This point of view has been influential, extending to proponents of other approaches, such as the actor-network theory. For instance, Callon and Latour, who are critical of *radical* locality in the sense that they allow for the fact that networks may have a certain context-transcending stability and extension, nevertheless state that:

> We cannot know who is big and who is small, who is hard and who is soft, who is hot and who is cold. (Callon and Latour, 1981, 296; see also Latour, 1987b, 229)

That is to say, even where a more extended stability exists, it is so contingent and precarious that it is not possible to base any argument or policy on it.[7]

In my view, the adjective "local" is one of the most unclear and abused notions within recent social studies of science and technology. The essential question, which is hardly ever asked, is: just how "local" is local? Certainly, emphasizing locality *is* justified when criticizing the facile universalistic claims made both by many philosophers of science and technology and by many social theorists. But it is plainly inadequate if locality is taken to imply the denial of any broader patterns or continuities in the development of science or technology—that is, if "local" is taken in a literal sense as pertaining exclusively to this or that laboratory or technology at such and such a moment. The existence and importance of such patterns or continuities can be easily established, in part even on the basis of those empirical studies that are so highly praised by the advocates of the locality thesis themselves. Some examples of such broader patterns are:

- the experimental nature of large parts of modern natural science (cf. Shapin and Schaffer, 1985);
- the black box or device feature of modern technology (cf. Latour, 1987b; Borgmann, 1984);
- the standardization of theoretical and experimental techniques and results (cf. Rouse, 1987, 111–119);
- the expansive character of many technological systems or networks (cf. Hughes, 1987; Latour, 1987b);
- the militarization of science and technology, especially in this century (cf. MacKenzie and Wajcman, 1985, part 4; Latour, 1987b, 167–173);
- the strongly increased importance of science to technology since the second half of the last century (cf. Schäfer, 1983).

Such patterns should not be seen as necessary, exceptionless regularities but rather as contingent, historical trends that require material and social work in order to be produced and maintained. Therefore, they are capable of being reproduced, strengthened, weakened, or changed. Moreover, since they always operate in specific contexts, they are not "independent variables" that would determine concrete scientific or technological practice from the outside, as it were.

Patterns and Their Normative Relevance

Let me first briefly expand on some of the above patterns. As we have seen in chapter 2, contrary to what is claimed by Collins (1985, 19), the norm of reproducibility does play a significant part in the practice of experimentation in the natural sciences. It also turned out that the normative use of the three types of reproducibility leads to nonlocal experimental knowledge. This is *not* to say that this knowledge is fully noncontextual and simply universally valid but rather that it has been *stabilized* against a number of specific variations in its local contexts of use. As many case studies show, applying the norm of reproducibility (in one or more of its forms) is not simply a matter of post hoc rationalization but an important feature of the experimental practice of the scientists themselves.[8]

The main reason why constructivists fail to see this is that they endorse the following implausible view: nonlocal norms (or rules, standards, methods, criteria, guidelines, and so on) should, if they are to be meaningful, *fully* determine scientific or technological practice, or else they are no more than retrospective rationalizations (by scientists, technocrats, or philosophers) of already established practice. Since this view is central to the locality thesis, let me document it somewhat further. Woolgar, for instance, states as a general conclusion that "logic and reasoning have a function quite different from that normally attributed to them. Far from compelling particular courses of action, they form the *post hoc* rationalization for ordered practices and conventional ways of proceeding" (Woolgar, 1988a, 48). And Wynne claims, referring specifically to technology, that "practices do not follow rules; rather rules follow evolving practices" (Wynne, 1988, 153). Finally, Richards and Schuster (1989, 700), in their criticism of Keller, ascribe to her (wrongly, I think) the belief that "*both* the standard and feminine methods ... really can command and constitute the scientific practices which are said to have been carried out under their direction." In contrast, they themselves see (nonlocal) methods as not literally efficacious, as floating above practice and as "myths which serve rhetorical and political purposes" (Richards and Schuster, 1989, 705).

I think, however, that it is more plausible to claim that nonlocal norms are practically effective *along with* (or as mediated by) all kinds of local factors. For example, the norm of experimental reproducibility was obviously consequential in such diverse contexts as that of Boyle, Faraday, and the gravity wave experimentalists, even if it did not determine the particular courses of action (see sections 2.3 and 2.4).

As a second illustration of nonlocal patterns, consider the role of standardized "fundamental laws" in theoretical physics. According to Cartwright (1983), such laws—for example, Maxwell's equations in electrodynamics or the Schrödinger equation in nonrelativistic quantum mechanics—fulfill the role of systematizing knowledge. As such they are strongly nonlocal. Yet, if we want to put these laws to work, they should be articulated with the help of a multitude of local models in order to make them empirically relevant and testable in different domains. Thus, Cartwright endorses a differentiated view of theoretical physics in which specific local models and systematic nonlocal laws are both necessary. Rouse however, in arguing for the essential locality of theoretical knowledge, highlights the models but underexposes the laws. For instance, when Cartwright (1983, 139) states that the aim of a fundamental law or theory is "to cover a wide variety of different phenomena with a small number of principles," Rouse (1987, 85) writes that Cartwright "presents us with a picture of theory as a disjoint collection of models."

A final example is from technology. In a detailed two-part paper, MacKenzie and Spinardi (1988a, 1988b) describe and analyze the postwar development of the U.S. fleet ballistic missile guidance and navigation technologies. With respect to the relation between technology and politics they write:

> we cannot simply assert an "interactive" view of technology and politics, point to the "seamless web" and leave it at that. For the seamless web is not shapeless. The task of a theory of technological change is to find its patterns and structures. (MacKenzie and Spinardi, 1988b, 612)

In a preliminary attempt, they identify two such patterns (MacKenzie and Spinardi, 1988b, 609–618). The first is that in as far as politics did systematically influence the developments, it was not so much "high politics" but rather "bureaucratic politics." As a consequence, the course of events cannot be understood in terms of top-down decisions and planning. Second, MacKenzie and Spinardi point to the black-box character of the U.S. fleet ballistic missile programs. A fairly stable boundary was created and maintained between internal "technical" and external "political" matters. As a matter of fact, this black-box pattern contributed much to the continuity and success of the programs.

My point is that the significance of patterns such as the above transcends by far any individual local situation in which they are embodied. Consequently, an adequate view of the development of science and technology should take into account *both* these broader patterns *and* the specific ways in which they are embodied—that is, reproduced or transformed—in local contexts. In other words, in concrete cases one finds not only change but also continuity.[9] Constructivist views, in stressing locality and contingency, tend to miss the importance of these continuities.

I have dwelt on this issue of locality at some length because it is crucial with respect to my theme. From a normative point of view, the significance of patterns and continuities can hardly be overemphasized. They provide the (always fallible) starting points for critical assessments of aspects of science and technology. For instance, if the development of nuclear weapons is completely unpatterned, what is the point of trying to get at least some democratic grip on it (for instance, by attempting to reduce the power of bureaucratic politics)? Or, to give another example, if the gender blindness of many sociologists of scientific knowledge is purely a local phenomenon, why criticize and try to change it? Why not just wait until some other idiosyncratic contingency produces a more favorable science and sociology of science tomorrow?

In sum, empirical studies of the development of science and technology show the existence and importance of broader patterns that normative social action requires in order to be sensible at all. In addition, it is important to note that this argument does not require a strongly realist interpretation of these patterns. In order to fulfill their role as premises for normative action, it is enough that they are considered plausible on the basis of the current criteria. In other words, it suffices to evaluate the claimed representational truth of such patterns from the perspective of our "natural ontological attitude."[10] Besides, in order to start normative reflexion or normatively inspired action, no appeal to universal norms is necessary. For this purpose, too, the existence of patterns of *shared* norms, or their construction by way of social negotiation or struggle, suffices. Something analogous holds with respect to the "normative relevance" of certain views or actions and to the "applicability of norms" in certain situations. Being normatively relevant is not a straightforward objective property of certain views or actions and whether a norm applies to a particular situation may be contested. Yet, in many cases *shared* beliefs about normative relevance or normative applicability exist.[11] Moreover, when such shared beliefs are still absent, one may try to bring them about through argumentative persuasion or social action.

5.5 ONTOLOGICAL, EPISTEMOLOGICAL, AND METHODOLOGICAL RELATIVISM

My next normative reflexion addresses the issue of relativism within constructivist approaches to science and technology. Relativism comes in different forms. Here I will distinguish between ontological, epistemological, and methodological relativism. Ontological relativism is the most radical position. It asserts that the world does not exist independently of human existence and human discourse. On the contrary, it claims that the world and its objects are no more than human constructs. Epistemological relativism remains agnostic with respect to the existence of a human-independent reality. But even if there were such a reality, it would, according to this view, be impossible in principle to obtain knowledge of it. Consequently, human knowledge claims cannot be divided into those that do and those that do not correspond to reality. In this respect all claims are on a par.[12] Methodological relativism is the most unassuming position. It merely claims that it is heuristically fruitful to study science and technology *as if* their objects are completely constituted by human, that is social, factors. That "reality" should not figure in explanations of the development of science and technology is an attractive methodological rule, because this leads to "interesting" empirical research.

Thus, *logically* methodological relativism does not entail epistemological and ontological relativism. Yet a study of the work of the most important proponents of relativism immediately reveals many argumentations and claims that amount to at least an epistemological relativist position. Adding epistemological relativism to the purely methodological position seems indeed natural and desirable: first, because it offers argumentative support to the methodology, which goes beyond pointing out its mere "fruitfulness" in claiming that it is adequate and in principle able to produce complete explanations of the phenomena under study; and, second, because, if methodological relativism is to have a wider significance than just for the empirical study of science and technology itself, it will have to be at least epistemologically underpinned. For example, without such an underpinning, the above-mentioned deconstructive critiques by social constructivists lack any ground. In practice, then, many constructivists turn out to endorse, at least, both epistemological and methodological relativism. Therefore, in the following discussions, I will leave the latter aside.

The *ontologically relativist* point of view can also be found among constructivists, although it is certainly not generally agreed upon. Particularly in the case of Woolgar, who writes within the more radical ethnographic tradition, ontological relativism plays a significant role. According to him, the organization of discourse *is* the object. Facts and objects in the world are inescapably textual constructions.[13]

But also occasionally in the work of other constructivists, ontologically relativist views are expressed. Thus, Bijker, Hughes, and Pinch (1987, 109) state that:

> In the social constructivist approach, the key point is not that the social is given any special status *behind* the natural; rather, it is claimed that there is nothing but the social: socially constructed natural phenomena, socially constructed social interests, socially constructed artifacts, and so on.

Thus, these views deny the natural (and, in Woolgar's case, also the social) world any independent existence and causal agency. A fortiori, realist interpretations of science and technology are claimed to be wrong.

There is, I think, a significant number of cases in which ontological relativism is an obstacle to critically assessing normative issues. These are cases where the phenomena concerned are not materially realized by scientific interference (for instance, in a laboratory) but where they result from causal intervention in a different setting. For instance, ontologically relativist constructivism casts a rather unusual and undesirable light on environmental action and politics. Consider an environmental issue such as "the hole in the ozone layer," and suppose that the claimed existence of this hole is, for some reason or other, seen as a problem. According to the ontological relativist point of view, this hole is identical to the discourse about it and it cannot possibly have any independent reality. Consequently the hole would simply disappear at the very moment we stopped discoursing about it, *even if*—and this is the crucial point from a normative perspective—we continued employing present technologies, such as aerosols, in an unaltered way! More generally, in all cases where we consider such kinds of issues as a problem, ontological relativism entails that "stop discoursing about them" constitutes a legitimate and effective *general* policy for solving them. But, although this would in many cases provide a relatively inexpensive solution to environmental problems,[14] it would obviously be a bad one. To be sure, such a type of solution might be conceivable in certain specific cases. But I wonder how many constructivists will really be prepared to endorse it as a general, practical policy?[15]

Next, let us turn to *epistemological relativism*. As mentioned above, this point of view is quite popular among present-day constructivists. Let me give just one illustration:

> think of one of those pictures which one constructs by joining numbered dots with pencil lines. Now imagine the world consisting of a

large sheet covered in almost infinitesimally small dots. The world is there in the form of the paper but mankind may put the numbers *wherever* he wishes and in this way can produce *any* picture.[16]

Epistemological relativism also gets into problems in the face of normative issues. To demonstrate this, I will consider the following argument by Collins. According to him, the sociology of scientific knowledge has consequences for science in the public arena. With respect to the use of forensic science in courtrooms, he writes:

> We should be looking to see if there are areas of forensic science which are more regularly challenged than others, and we should be asking if unchallenged forensic evidence which is taken to justify a prison sentence would be taken to justify a radical conclusion *within* science. (Collins, 1983, 99)

Now, of course, sociologists of scientific knowledge may ask questions like these. But since for them "to justify" means nothing more than "to accept as a result of contingent, social negotiations," no normatively relevant conclusions can be drawn from answering these questions. If Collins really sticks to the relativist assumptions of his program, the only conclusion that *is* justified is that the negotiations within science and within the courtroom are different, period. Consider, for instance, the claim made within a forensic context that "the accused has blood group A." Then the epistemological relativist framework provides no clue at all as to why referring to apparently similar claims and procedures in a scientific context ought to be relevant to accepting or criticizing this claim.

An Alternative: Nonrepresentational, Referential Realism

In cases like the above, constructivist relativism has unacceptable normative consequences. In order to avoid these consequences we need a modest form of realism concerning science and technology, instead of ontological and/or epistemological relativism. Elsewhere I have developed a "referential realism" that, on the one hand, takes into account the independent existence and the causal agency of reality and explains when terms from the description of experiments and technologies (for example, "the hole in the ozone layer" or "blood group A") may refer to elements of this reality. The ontological claims of independence and causal agency imply that objects, such as that referred to as "the hole in the ozone layer," cannot be made to disappear by stopping discourse; and the epistemological explanation of reference provides a necessary (and hence a normatively relevant) condition for the acceptability of claims such as "person P has blood group A," in both a scientific and a forensic context. On the other hand, this form of realism is explicitly nonrepresentational, and therefore compatible with the interpretative flexibility of our conceptual interpretations of nature, which is rightly stressed in constructivist work. Thus, referential realism involves a conceptual but not an epistemological or ontological relativism.[17]

Such a referential realism is a precondition for making sense of, among other things, environmental controversies and proposed policies. It does not provide a recipe for closing such controversies or for choosing a unique course of action by appealing to "the one and only real representation of nature." Neither does it exclude attempts at criticizing any *specific* claims concerning the reference of terms or the acceptability of statements. Critics may argue that a term in question does, in fact, not satisfy the criterion of reference. Or they may claim that, although the term does refer, its conceptual representation is implausible. In Radder (1988) and in the preceding chapter, I have shown the *feasibility* of such a form of realism, in particular in view of the problems "change" and "work." The present section argues for its *desirability*, in order to be able to make sense of (normatively inspired) action. Since, in my view, philosophical "proof" just consists of showing both feasibility and desirability, my conclusion can be that referential realism is a more plausible philosophical interpretation of natural science and technology than constructivist relativism.[18]

I should like to conclude this section with a short parable that illustrates the points made. It is inspired by MacKenzie's (1989) paper "From Kwajalein to Armageddon? Testing and the Social Construction of Missile Accuracy." In this paper, MacKenzie offers an interesting account of the way in which the interpretative flexibility of the concept of missile accuracy was restricted and a social closure reached. However, in concentrating on the subtitle of his paper, he tends to miss the issue implied in its title—namely, the possibility of a complete disappearance of the human species as a consequence of a worldwide nuclear war. Imagine that, after Armageddon, a number of scientists from outer space visit the earth. After their return a scientific controversy develops among the extraterrestrials concerning the question of whether humanity died from, was killed by, was murdered by, or was massacred by nuclear missiles.[19] Indeed, there is much interpretative flexibility here! Yet, the moral of this story is that all these descriptions have one element in common: they all refer to the same material realization. And, of course, it is just this common referent, this potential consequence of the use of nuclear weapons technology, that is all-important in a normative reflexion on this technology.

5.6 THE ACTOR-NETWORK THEORY

My final subject of normative reflexion is the actor-network theory. In this theory, proposed by Callon, Latour, and Law, science and technology are seen as goal directed—that is to say, directed toward stability, control, and extension of actor-networks (see Callon, 1987; Latour, 1983, 1987b; Law, 1987). Yet this stability, when reached, is never secure. It is always provisional and precarious because of the fact that it may be undone by antagonistic actors at any time. The actors in a network include nonhumans as well as humans—not only the Renault company, Pasteur, and the Portuguese colonialists but also electric cars, microbes, and oceanic wave currents. The elements and interactions within these networks form a heterogeneous but seamless web.[20] Therefore, according to this theory, it is impossible to isolate and reify nature, society, science, or technology as separate entities performing specific types of activity in the production of facts and artifacts. On the contrary, the existence of these "entities" is in fact the *result*, and not the cause, of network interactions. Consequently, according to the proponents of this approach, the notions of "nature," "society," "science," and "technology" can never be used to explain the development of the production of facts and artifacts. Instead, it is their reification that is in need of explanation.

The most conspicuous normative relevance of the actor-network theory lies in its radical deconstructive character. Its criticism is not only directed at "internal" scientific or technological determinism but also at the sociological reductionism of social constructivists. Since nonhuman actors do play significant roles in the networks, a reduction to the social is impossible. So, to a certain extent, this theory is more satisfactory than social constructivism. In this approach the hole in the ozone layer may be a relevant and powerful element of a network of environmental actors.

Yet, since it is simultaneously claimed that this hole—and more generally, nature—is merely the result of the closure of a controversy among actors in a network, an important ambiguity remains. Indeed, this claim immediately invites the question of how it is possible that an actor *both* acts during the process of network interactions *and* comes into being only as a product of these interactions when the controversies about its existence have settled. Callon and Latour, in particular, deny the relevance of the first part of this question. As we shall see in more detail below, their view implies that actors can be said to act only in so far as they turn out to be on the winning side in the settlement of the controversies. From a conceptual point of view, such a claim is quite puzzling, to say the least. But I will argue below that it is also undesirable from a normative point of view.

Law's view, which seems to differ somewhat from that of Callon and Latour with respect to this issue, is more satisfactory. He proposes that actors in the natural world, as perceived by the system builders, may constitute part of the *explanans*, as long as nature is not given a privileged or exclusive status in comparison to the other actors in the network.

> Depending, of course, on the contingent circumstances, the natural world and artifacts may enter the account as an *explanans*. And in case it is thought that I am giving too much away to realism, let me say that, so long as we are concerned exclusively with networks that are being built by people, then "nature" reveals its obduracy in a way that is relevant only to the network when it is registered by the system builders. (Law, 1987, 131)

Consequently, according to Law, nature is not purely the result of network interactions, even if "revealing nature" requires the activity of system builders. Reformulated in this way, this view will, I think, be acceptable to many students of science and technology. It may, for instance, be interpreted in a somewhat different language as saying that the ascription of causal links is always theory laden and that in any real situation a multiplicity of (heterogeneous) causal factors is operative. Such a view would be compatible with the referential realism discussed in the previous section.

The Winner's Point of View

There are a number of aspects of the actor-network theory that are especially problematic with respect to normative matters. One such aspect concerns its notion of goal orientation—that is, the tendency towards stability, towards black boxing, towards control. Callon and Latour (1981, 283) write about this:

> in order to stabilize society everyone ... needs to bring into play associations *that last longer than the interactions that formed them.*

What is the nature of these stabilization processes? A principled answer from the viewpoint of the actor-network theory would seem to be that stabilization, or order, is the contingent result of an agonistic struggle between heterogeneous actors, some of which turn out to be winners and others losers. However, within the actor-network approach, there is also a tendency or danger to analyze the process of network or system building exclusively from the point of view of the winners, the successful actors, whoever they are.[21] Put differently, the idea that there are no *guaranteed* winners implies *both* that in principle there are always a number of potential winners *and* that there are always winners and losers. Actor-network theorists tend to stress the former and forget the latter.

I observed earlier that the tendency to emphasize the winning side is intrinsically related to the metaphysical assumptions of the actor-network theory. Here are some other, more direct illustrations of this tendency. First we can see that this bias is quite apparent in the definition of the notion of an actor proposed by Callon and Latour (1981, 286):

> What is an "actor"? Any element which bends space around itself, makes other elements dependent upon itself and translates their will into a language of its own.

Clearly, if we are to take this definition seriously, there are only winning actors! Another quotation from this article is also revealing in this respect:

> In the . . . conflicts we have just described, there are indeed winners and losers—at least for a while. The only interest of our method is that it enables these variations to be measured and the *winners* to be designated. (Callon and Latour, 1981, 292, emphasis added)

Furthermore, when no observable battle is going on any more, Latour advises us to accept the status quo. Students of science and technology should *follow* technoscientists, not ask them critical questions (see Latour, 1987b, 100). Second, in all the leading metaphors of the theory, the agonistic field appears from the perspective of the winners. We read about system builders, Princes, Leviathans. But what about the losers? Who are their representatives at the metaphorical level? On this question the proponents of the actor-network approach are conspicuously silent. Third, some case studies reveal the same tendency. For example, from Law's account of Portuguese colonial expansion we learn a lot about the (ultimately) successful actors but hardly anything about the perspectives of the colonized inhabitants of Africa or India.[22]

The general purport of my criticism here is that a more symmetrical treatment of *all* the relevant actor perspectives is desirable. This issue is extremely important with respect to normative questions. At some point, all liberating or emancipatory movements (whether of ethnic minorities, of women, or of colonized third-world citizens) have acknowledged the importance of writing *their* history and of accounting for its absence in the "official" history of the winning sides. Since technoscience has been inextricably involved in these histories, the same requirements of symmetry should be put on its study. Paraphrasing the quotation from Latour given at the beginning of this chapter: pursuing democracy, or "the People becoming the Prince," is not possible if the People are not allowed a voice of their own.

What is Technoscientific Success?

Next, consider the notion of "success." According to the actor-network theory, the longer the spatiotemporal extension of a network, the more difficult it is to break it up and, given the goal of technoscience, the more successful the technoscientists are. However, "length" as a criterion of technoscientific success does not seem to be appropriate. An obvious reason for this is the fact that a long chain is not necessarily strong. After all, a chain is as strong as its weakest link, as the saying goes. Many authors have taken this fact into account in their theories of science and technology. Kuhn (1970b), for instance, distinguishes "puzzles," which are relatively unproblematic problems, from "anomalies," the real weak points within scientific paradigms. And with regard to technology, Hughes (1987) points to "reverse salients" and "critical problems" as focal points for improving the strength of technological systems. The same point can be put somewhat differently. The criterion of length suggests that assessing success in technoscience is merely a matter of counting enrolled actors: "numbers, more numbers" (Latour, 1987b, 60-61). However, such a quasi-objectivist account of technoscientific success will not do. Since, in practice, all scientific theories and technological artifacts have to contend with difficulties, the crucial question is how serious they are taken to be. Deciding this question is not only a matter of counting but also one of judging.

Presumably because they sense the problematic character of the notion of length of networks, Callon and Latour at times also refer to their strength (Callon and Latour, 1981, 292; Callon, 1987, 93; Latour, 1987b, 33–34, 121–124 and 202–203). However, their general point of view entails that strength can never be a causal factor; it is always a result of a (successful) development of a network. To be successful *is* to be strong. In other words, since the notion of strength cannot be independently specified within the actor-network theory, it cannot function as an explanation or criterion of technoscientific success.

The absence of any specification of technoscientific success (and thus of failure) is important with respect to normative questions. For instance, if we want to decide whether we should develop or use a certain technology, we necessarily have to claim in advance something about what it means to have the technology work. If we can say nothing at all about the future, as is implied by the actor-network theory, any choice or policy is no more than a lottery. That is to say, if we consistently endorse Latour's (1987b, 258–259) rules and principles of the actor-network theory (and not just pick out distinct pieces), the theory turns out to be useless in constructive attempts aimed at evaluating the advantages and disadvantages of proposed technologies.

In contrast, I think that the social study of science and technology should overcome this fear of becoming explicitly normative. The same concern is phrased by Fuller (1988, ix) as follows:

> If sociologists and other students of actual practice wish their work to have the more general significance that it deserves, then they should practice some "naturalistic epistemology" and welcome the opportu-

nity to extrapolate from *is* to *ought*. If these empiricists realized, following Max Weber, that the inferential leap from facts to values is no greater than the leap from our knowledge of the present to our knowledge of the future (a leap that the empiricists would risk in the normal course of their inquiries), they would be relieved of the peculiar combination of fear and loathing which normally prevents them from encroaching on the philosopher's traditional terrain.

In order to become relevant to the evaluation and direction of science and technology, we should first of all actively look for patterns of scientific and technological success, and systematically analyze the conditions of their applicability to find out what is implied and required in their extrapolation. Our response to the fact that prediction and control in science and technology is essentially uncertain should not be that we do not make any deliberate choices or policies at all. Instead, a wiser strategy seems to be to try to implement such scientific or technological projects that, according to the best of our knowledge, would cause minimal damage should they fail, and the material and social realization of which is democratically supported by all the people involved. Let me briefly describe two examples of approaches that follow such a strategy.

First consider the analysis of technologies as *attempts* at realizing closed systems.[23] Here "closedness" is claimed to be a necessary condition for the stability and reproducibility of a system. If we succeed in producing and maintaining a stable and reproducible technological system, we are in control of the situation and can use it for the purposes at hand. Thus, closedness is related to matters of safety, risk, and other social issues concerning technology. But in order to reach such a situation of control we should, among other things, be able to fulfill certain material and social conditions. Therefore, this analysis, when applied to particular cases, can be used when normatively assessing technologies in a conditional way: *if* we want to have technological success, we will have to close the technological system in question and thus have to satisfy such-and-such material and social conditions. The question we then must confront is whether the realization of these conditions is and will remain practically feasible and normatively desirable.

Perrow's (1984) analysis of "normal accidents" provides a second illustration. He gives a systematic account of high-risk technologies. It is based on two theoretical notions—namely, *complexity* of systems, implying that unanticipated events may and will happen, and *tight coupling* within systems, implying that failure in one part of the system will rapidly spread to other parts. Perrow's claim is that in complex and tightly coupled technological systems (for example, nuclear power stations, chemical plants, biotechnological systems, aircraft), major accidents can be expected to occur "normally." On the basis of his analysis and his many examples, he then discusses the normative question of what is to be done when we take into account not only economic and technical but also social and cultural values and norms.

In the context of the present discussion, it is not the particular content of these two approaches that is at issue but rather the type of argument they represent. Thus, whatever our opinion of the specific merits of these approaches, they are at least attempts to ascertain, analyze, and evaluate certain normative criteria of technoscientific success and failure. Because of their normativity, these theoretical analyses may possibly be helpful in assessing the relative acceptability of technologies. As such, they clearly contrast with the actor-network theory.

5.7 CONCLUSION: ANALYTICAL, CRITICAL, AND CONSTRUCTIVE REFLEXIVITY

At the beginning of this chapter, I introduced a distinction between analytical, critical, and constructive normative reflexivity. In this final section, I will summarize my reflexions on constructivism in these terms. So, let me start with the question of what are the main *analytical* results. First, it is notable that the different views within constructivism display both overlapping themes and relevant differences. The normative or normatively relevant presuppositions, assumptions, and implications of constructivist approaches to science and technology, though not unpatterned, are many and differentiated. They can be found in systematically underresearched subjects (the military), in the description of specific cases (focus on winners), in the theoretical terms (locality issue) or in the philosophical interpretations of an approach (relativism). Consequently, normative reflexion on (the study of) science and technology should neither be restricted to one subject, such as "risk" or "gender," nor depend exclusively on one concept, such as "instrumental reason" or "the ideology of representation."

The above also shows that empirically factual and normative issues are fused. With Pels (1990), I would like to call this the "natural proximity" of facts and norms. It entails, first, that analytically the distinction between facts and norms is sensible and fruitful; second, that no primacy, either of facts or of norms, should be posited; and, third, that any claimed reduction of norms to facts or the reverse is illusory. Indeed, many of the analytical reflexions presented here aim at making explicit the normative issues that naturally go with the constructivists' facts.

Of course analytical, critical, and constructive reflexions build on each other and are thus not unconnected activities. Therefore, just as with the analytical, the *critical* normative reflexions in this chapter are many and differentiated. I have pointed out that there is not just one normative problem of the constructivist study of science and technology and that any particular problem is not always a problem of all the approaches within constructivism. Consequently, the one and only "master solution" to the normative problems of (the study of) our technoscientific world does not exist.

At the most general level, my critique concerns the nearly total absence of normative reflexion within constructivist studies of science and technology. This also includes arguing against the legitimations of this absence—for instance, by means of a criticism of the empiricist claim that constructivists do no more than simply "follow" technoscientists and register what they say, write, or do. To be sure, I do not claim that individual proponents of constructivism may not endorse all kinds of normative views (see the examples in section 5.2) or that they may not be involved in all kinds of normatively inspired activities. Rather, the nature of the constructivist approaches implies what I would like to call (after a Dutch nursery rhyme) a *"Kortjakje* effect": at work analysts claim to be academic, "disinterested" researchers trying to find out how science and technology really operate, while normative preferences are apparently only allowed to play a role on Sunday, be it in church or in the pub. In fact, as I have tried to show in this study, such an artificial separation between cognitive and normative views is not only undesirable but also impossible. Students of science and technology are themselves part of a complex web, in which cognitive and normative aspects are intertwined.

Second, I have criticized certain constructivist doctrines—about locality, relativism, and technoscientific success—that proved to be obstacles for the (practically unavoidable) normative assessments and actions relating to the direction and impact of science and technology. Making choices more democratically and more justly, anticipating possibly hazardous or undesirable consequences, and critically assessing the role and meaning of technoscience remain important regulative ideals, even though (or better: especially when) we rightly do not believe in the scientistic ideal of straightforward historical progress through science and technology, or in the modernist ideal of the full controllability of society.

Third, I have been more explicitly normative myself in arguing that missing particular issues, such as "the role of gender" or "the loser's perspective," will be disadvantageous for emancipatory movements. But even someone who does not share the values of these movements can take a point here. After all, it is certainly the case that "gender" or "losers" concern prima facie obvious and common aspects of technoscience, which can only be ignored at the price of getting an incomplete or biased picture of it.

Let me conclude with some observations concerning *constructive* normative reflexions—that is, alternative proposals that are more adequate with respect to normative issues. The proposals discussed in this chapter concerned the significance of nonlocal patterns, such as the role of bureaucratic politics in nuclear missile technology, the desirability of a modest form of realism (namely, referential realism), and the policy relevance of success criteria, for instance Perrow's theory of "normal accidents." These alternatives do not imply a return to the often simplistic, universal schemes that have been put forward so many times by philosophers of science and technology.[24] As I have stressed, the alternatives should not only be normatively desirable, they should also be empirically adequate (in the case of patterns), feasible (in the case of referential realism), and workable (in the case of success criteria). Put differently, due to their normative commitments, full-fledged normative reflexions should maintain a critical distance from the world of science and technology. In particular, they should not go along with Latour's postcritical call: "Down with Kant! Down with the Critique! Let us go back to the world . . ." (Latour, 1988, 173). At the same time, however, in trying to live up to standards of empirical adequacy, feasibility, and workability, they should avoid being trapped into a hypercritical, but ultimately suicidal, constitutive reflexivity.

CHAPTER 6

Experiment, Technology, and the Intrinsic Connection between Knowledge and Power

6.1 INTRODUCTION

A major aim of the preceding chapter was to remove certain "constructivist obstacles" for developing a normatively more satisfactory study of science and technology. In the present and subsequent chapters, I will attempt to proceed in a more positive vein by proposing an approach that enables us to make concrete, normatively relevant analyses and assessments of science and technology. First I will link up with the discussion of technoscientific success in section 5.6 and, more particularly, with the notion of closedness as a necessary condition for the stability and reproducibility of experimental or technological systems. Thus, in this chapter both science and technology will be analyzed as "attempts at the realization of closed systems."[1] The use of this substantial analogy makes it possible to explicate both cognitive and social similarities and dissimilarities between science and technology.[2]

The notion of prediction can be used for presenting the general line of my argument. As is well known, prediction in the natural sciences is always conditional: "under conditions c, cause x produces effect y." The investigation of the proper conditions is, in particular, the task of the experimenter and the technician. If we only pay attention to theories, to reasoning, and so on, the principal concerns are x and y and their relations. However, if we also take into account the realization of knowledge in experimentation and technological production, different issues come to the fore. In particular, successful prediction can be seen to presuppose the possibility of a material and social intervention in, and control of, the experimental or technological system and its environment; moreover, it becomes clear that this intervention and control not only require scientific or technological expertise but also social power. In other words, in this way we run up against intrinsic relations between cognitive and (especially macro-) social aspects of science and technology. The main objective of this chapter is to provide an analytical model for investigating such relations. The analysis in terms of the production and maintenance of closed systems makes it possible to connect, in a direct and precise manner, experimental or technological success and macrosocial factors.[3] Furthermore, this model is both sufficiently general to be fruitfully applied to a number of cases, and sufficiently flexible to do justice to the differences between these cases.

The general features of the approach are described in section 6.2, in which the notion of closed (experimental or technological) systems is analyzed. In the remainder of the chapter, this notion is developed and used in two different ways. In section 6.3, I demonstrate its practical applicability in two cases. The first deals with particular aspects of nuclear energy technology, the other with some entomological tests of insect eradication techniques. In discussing these cases, I will also make some comments on other theoretical views concerning the relations between science and technology. Next, in section 6.4, the model is used to analyze and assess public debate on technology and technology policy. By comparing it with the conception of science and technology that prevails in public discussions and policymaking, the present approach turns out to imply certain normatively relevant consequences. The points are made by means of an analysis of the so-called Brede Maatschappelijke Diskussie, or BMD (Extensive Social Discussion), on energy policy, which took place in the Netherlands during the years 1981 to 1983. Finally, in section 6.5, it is concluded that scientific or technological knowledge and social power are intrinsically related, and the nature of this relationship is briefly explored.

6.2 THE PRODUCTION AND MAINTENANCE OF CLOSED SYSTEMS

In chapter 2 I mentioned closedness as one of the aspects of (successful) experimentation. Often, closedness is taken to imply roughly the following. We distinguish between system and environment ("inside" and "outside") and call a system "closed," when the inside is not influenced by the outside. A crucial point is, of course, exactly what we mean by the notions of "system" and "influence." With respect to the former, I will limit myself for the time being to the following characterization: a system is "a whole of mutually interacting entities at a certain spatiotemporal location." As for the latter, in an ontological conception, closedness is conceived as the isolation, whether or not approximate, of such a system in reality with respect to *any* conceivable influence (see, e.g., Bhaskar, 1978). Such an ontological conception, however, is not suited to analyzing practical experiments and technologies. It has to be relativized in two ways.

First, in the assessment of closedness, we are never dealing with experimental or technological systems "as such," but with systems as described or interpreted by certain theories. These theories specify (and hence restrict) the number of types of possible influences upon the system by the environment. Therefore, closedness is always relative to the scientific or technological description that is chosen. A further relativization has to do with the fact that an experimental or technological project is always a complex process that, by being interpreted theoretically, is split into separate experimental or technological situations. In view of a successful realization of the project, it is not necessary (and in general it is impossible) to predict or control the development of all theoretically interpreted situations. We have to confine ourselves to those situations that are *relevant* in view of the *problem* and *aim* of the project.

Moreover, the provisional formulation of the idea of closedness given above should not only be relativized but also extended. Typically, closedness requires not merely that there be no (relevant) influence of the "outside" upon the "inside" but also the reverse: certain effects of the system upon the environment must also be eliminated or controlled. That is to say, in general, closedness requires that there be no (relevant) *inter*action or *mutual* influence.

In the light of these considerations, I propose to call a system S, as described by a theory T, *closed* in a certain time interval if and only if the following conditions apply:

1. the relevant situations that actually occur within S do not have any sufficient conditions outside S;

2. those conditions outside S are fulfilled, which are necessary for the occurrence of those situations inside S that enable us to realize the aim of the (experimental or technological) project in a stable and reproducible manner;

3. the situations that actually occur within S are not sufficient conditions for the relevant situations outside S.

The underlying idea is that, if one wants to be able to predict and control experimental or technological processes, one should require stability and reproducibility of the relevant situations within S and exclude undesirable effects by S upon its environment.[4] In order to attain this state of affairs, it is necessary to close the system by eliminating (or preventing the occurrence of) the sufficient conditions mentioned in the definition above and by producing (or maintaining) the named necessary conditions.

Up to this point I have focussed on the theoretical analysis of the production and maintenance of closed systems. But, of course, this forms only one side of the problem. Experiments and technological projects do not need only to be theoretically described: they also need to be realized in practice. Closed systems are nearly never "just found." We practically always have to produce and maintain them through active intervention. The crucial point is that, to this end, theoretical-scientific or theoretical-technological knowledge plus experimental or technological know-how are surely necessary, but they are not sufficient. In order to be able to guarantee the required control of the relevant situations we need, at the same time, knowledge and possibilities of control of the *social* setting of the system.

On the basis of the above definition of closed systems, the coupling of cognitive and social aspects can be resolved into three components:

1. the sufficient conditions that would disturb the "free" evolution of the system or would produce undesirable effects outside the system have to be eliminated or precluded from taking effect by means of the appropriate social measures;
2. the required necessary conditions have to be materially and socially realized;
3. assessing the plausibility of the theoretical description and the relevance of experimental or technological situations proves to be not just a cognitive but also a social issue.

Basically, the specific notion of closedness that I will use in this chapter means: "closedness with respect to (theoretically, materially, and socially) undesirable influences and disturbances." Thus, this notion is perfectly compatible with the fact that, according to the different, ontological terminology mentioned above, experimental and technological systems are "open" systems, since they will always maintain some interactions with their environment.

Still another definition of closed (and open) systems is submitted by Pickering:

> A closed system is one which is perfectly well understood, measurements upon which yield observational facts commanding universal assent. An open system is one imperfectly understood, measurements upon which are open to an infinite variety of interpretations. (Pickering, 1981, 218)

However, as a definition and/or criterion of closedness such a description is not specific enough. This becomes particularly clear from the case analyzed by Pickering, the experimental quark controversy. The problems with the interpretation of the experiments in question (for example, those concerning the homogeneity of the electric fields) were related to measurement interactions *within*, and not so much to closedness of, the experimental systems. Pickering's account shows, however, that experimentation (and the same holds good for technological production) includes more than the production of closed systems. In the present chapter, I exploit the notion of closedness because it appears to be particularly suited for a comparison of experiment and technology that focuses on *macro*social aspects.

The above analysis in terms of closed systems aims to integrate a system approach to the problem of science and technology with an actor approach (cf. Law, 1984). First, the system is defined in a rather restricted way—namely, with respect to scientific or technological problems and goals. This has an analytical advantage in that it provides a clear and well-defined starting point. Next, by taking into account the notions of plausibility and relevance and the (material and social) conditions for scientific or technological success, one contextualizes the system and relates it to the different perspectives of social actors. Because the context may be chosen to be as wide as is needed for the problem at hand, one gets the additional advantage of potential comprehensiveness.

6.3 THE RELATION BETWEEN EXPERIMENTATION AND TECHNOLOGICAL PRODUCTION

So far the analysis has been rather formal. I have argued that closedness is a necessary condition of experimental or technological success, but as yet *nothing* has been said about the feasibility and desirability of realizing closed systems. The question, then, is to what extent we are in practice able and willing to fulfill the requirements for closedness outlined in the preceding section. By discussing this question for laboratory experiments and for technological projects successively, the analogy between experimentation and technological production can be made explicit.

In laboratory contexts, closing experimental systems often proves successful. Commonly, scientific experiments take place on a relatively small scale. The smallness of scale is not only spatial, as Latour (1983) shows in his study of Pasteur, but also concerns the time dimension: because in experimentation, the cognitive aspect is more to the fore—in comparison with technological production—experimental systems, in general, need only be kept closed during a relatively short time (cf. Hacking, 1983a, 39). Hence, the social control that is needed for successful experimentation may seem to be trivial. This, however, is merely a consequence of the fact that in our society, experimental natural science is an established and institutionalized phenomenon. That it is permitted to forbid playing children or antiscience demonstrators from disturbing the order produced in the laboratory seems natural to us. But in a society in which natural scientific experimentation is illegitimate or illegal (as once magic and alchemistic practices were), things will come about rather differently.

In this chapter, I also analyze technology in terms of the production and maintenance of closed systems. Thus, I endorse the view that a formal similarity obtains between technological production and experimentation. In order to make both experimental and technological systems work, we must try to close them.[5] This similarity, however, remains a formal one. The actual use of experimental results in technological projects may lead to substantial differences between the experimental and technological realizations. This applies in particular to those technological projects for which, owing to social demands, long-term and large-scale operation is required.

With the help of the theoretical analyses from the preceding section, the differences in degree between experimentation and technological production can be further explained. I will illustrate these differences, including the new and unexpected problems they often entail, mainly with the help of two examples: one taken from nuclear energy technology, the other from the technological uses of entomology.

Closing Nuclear Energy Systems

First, I observed that the problems with the control of the necessary and sufficient conditions of closedness may be considerably greater for the technological systems than for the corresponding experimental systems. This clearly applies to the case of nuclear energy. For example, it is often concluded that high-level radioactive waste can be safely enclosed in borosilicate glass containers and subsequently stored for a long time in subterranean salt domes. This conclusion, however, has been questioned, because it has been reached on the basis of laboratory experiments, where the glazing of the waste can take place under "ideal" circumstances (see Ringwood and Willis, 1984). In industrial applications, the melting and cooling of the glass is slower and less regular. As a consequence, the glass always contains tiny cracks that have the effect that, under the influence of the water in the salt domes and of the radioactive radiation from the waste, the glass may corrode much faster than might be expected in the light of the laboratory experiments. This argument, in other words, makes it doubtful whether we really will be able to maintain the material conditions for the closedness of the technological system "waste in salt domes" throughout the required long period.

Second, in general the social conditions necessary for a permanent closedness of technological systems, especially of large ones, can be expected to be much more difficult to create and maintain than in the comparable experimental case. For instance, nuclear energy, being a large-scale and complicated technology, demands extensive interventions from the central government. Guaranteeing the safety of the enrichment and reprocessing plants, of the nuclear power plants themselves, and of the stored radioactive waste requires all kinds of political measures: increased secrecy, intrusions into the privacy of workers and neighbors, restrictions on the freedom of demonstration, and the like.[6] According to our present knowledge of society, such centralist and antidemocratic measures are needed in order to guarantee a permanent closedness of the system of nuclear energy production. Whether these measures are, and will remain, feasible and desirable is of course quite a different matter.[7]

Third, there is the problem of the appropriate theoretical description of the system and of the relevant situations for which closedness is required. Concerning this issue, the differences between technology and experimentation will often (but not always: see the entomology case, later on) stem from differing impacts of the systems upon their environment. Although laboratory experiments (for example, recombinant DNA experiments) may also have undesirable effects upon their environment, the effects of technologies will in general be more diverse and far-reaching and hence more difficult to ascertain and to control.

For instance, is the routine escape of low doses of ionizing radiation a relevant issue with respect to the closedness of a normally functioning nuclear power plant? According to the BMD Steering Committee, it is not. They state that during normal operation, the amount of leaking radiation is small compared to the radiation originating from other sources. The implicit conclusion is that normal radiation leakage is not a relevant issue for the closedness of the technological system in question. In this respect, normally operating nuclear power plants are considered to be safe (see Stuurgroep, 1983, 84–87, and 1984, 207 and 351). But according to others, low radiation doses may be positively relevant to the closedness and safety of nuclear power plants (see Biesiot, 1983; Reijnders, 1984; Nussbaum, 1985). In particular, they dispute two things: the manner in which dose-effect relationships are extrapolated toward the domain of low doses; and the statistical methods that mainly work with averages over the *whole* population. With regard to the latter issue, the critics point, for instance, to the greater risk that specific groups of the population (such as radiological workers or young children) may run. The account of this controversy illustrates well the relationship of cognitive and social aspects: calculations in terms of "risk acceptable in the average" (and, thus, *not* in terms of "risk acceptable for specific groups") are simultaneously cognitive and social.

The BMD Steering Committee tries to resolve the problem by asserting that "the effects of radiation are better known than those of nearly all other injurious substances" (Stuurgroep, 1983, 87). This is a rather curious statement, especially considering the fact that the Steering Committee itself notices the existence of divergent points of view on the issue (Stuurgroep, 1983, 85). By passing over the contents of these viewpoints, the committee pursues a strategy that aims to define the topic of "low doses" as being irrelevant to the production and maintenance of closed nuclear energy systems. That is to say: in this particular case they attempt to close the *system* by closing the *debate* (about certain of its aspects). As will be clear, this possibility is a direct consequence of the nonontological definition of closedness given in the previous section.

The Boll Weevil Eradication Experiment

To demonstrate the wider applicability of the theoretical viewpoint advocated in this chapter, I will deal with another example, taken from a different field—namely, entomology. For this purpose I make use of Perkins's (1982) study of a "Pilot Boll Weevil Eradication Experiment," or "PBWEE" (cf. also Groenewegen, 1985). Perkins offers an illuminating account of the problems that arose in this large-scale (an area with a radius of twenty-five miles) and long-term (two years) field experiment. It was designed to test entomological techniques for preventing damage to the cotton harvest caused by boll weevils (in the United States, between 1971 and 1973). Thus, the analysis in terms of closedness proves to be applicable not only to technological production (as in the case of nuclear energy) but also to technological testing in field experiments.

The initial problem of the experiment was how the technological system (that is, the cotton area) might be radically freed of boll weevils. However, in spite of the application of eight different methods, successful laboratory techniques could not be reproduced in the field. This failure was caused by all kinds of difficulties. For instance, the migration of boll weevils from outside to the area inside could not be prevented; that is to say, it was not possible to realize materially one of the necessary conditions of closedness. As a consequence, it turned out to be impossible to answer the question for which the experiment had been designed—namely, the question of whether entomological technology was able to eradicate the boll weevil from the test area in a reproducible manner (Perkins, 1982, 130).

The example also provides a clear illustration of the fact that social conditions are essential to achieving or not achieving closedness in technological projects. Thus, Perkins points to several problems that arose in performing and evaluating the experiment:

> A prime requirement of the PBWEE was that every farmer in the experimental zone co-operate with the goals and procedures involved; one grower who did not could jeopardize the entire effort by "raising" boll weevils on his improperly treated cotton.

However:

> In fact, many of the farmers of the area apparently grew the cotton solely to collect the government price support payments and never even bothered to harvest their crop, let alone exercise careful insect control practices on it. (Perkins, 1982, 129)

Another problem was the following. One of the suppressive techniques consisted of the release of large numbers of reared sterile male boll weevils, which was meant to block the reproduction process in the field. The efficiency of this technique was also lowered by difficulties that had to do with the realization of its social conditions:

> The quality of the "sterile males" declined even more as the termination of the experiment on August 8 approached, and temporary personnel in the mass-rearing facility began to lose morale due to impending layoffs. (Perkins, 1982, 137)

In both cases (that of the farmers and that of the temporary personnel), a social condition that was necessary for a good performance and an unambiguous evaluation of the experiment turned out to be unrealizable.[8] Conversely, we can conclude that, if one wants to make the eradication technology work, one has to be able, materially and socially, to realize the entomological knowledge in the open field. To this end, the availability of laboratory knowledge does not suffice: one also needs the relevant social power.

Finally, there is the question of the correct theoretical description and of the relevant situations of the system. On this point, too, clear differences may appear between laboratory and field experiment. In the PBWEE, these differences entailed a change in the laboratory criterion of closedness. In the course of the experiment, the above-mentioned material and social problems led to a proposal to redefine the notion of "eradication": eradication would have to be conceived as "elimination of boll weevils as an economic pest," instead of "total and permanent absence of boll weevils," as was usual in the laboratory experiments. As a consequence, the criterion of closedness changed as well. For instance, migration of a *restricted* number of boll weevils from outside into the test area became irrelevant to achieving closedness of the system.[9]

The Science-Technology Analogy

By way of conclusion of this section, I will comment briefly upon two other positions in technology studies, both of which make use of analogies between science and technology. MacKenzie (1989) considers the connection between cognitive and social aspects of technological testing from the point of view of the sociology of scientific knowledge. More in particular, he offers an account of the testing of the accuracy of nuclear intercontinental ballistic missiles by the U.S. Air Force. In order to measure the precise points of impact of the warheads, a network of microphones, spread out over a certain area in the Kwajalein lagoon, is used. Such a detection instrument can only work, however, if it is not disturbed, say, by the fishing boats of the Marshallese, the original inhabitants of the islands. Thus, in this case, one condition of successful technological testing is that these people are deported elsewhere or, at least, that they are systematically excluded from the test area. This condition has indeed been realized. But note that this sort of condition cannot be fulfilled everywhere. As MacKenzie observes, the deportation of Americans or Canadians for the purpose of testing missile accuracy along a north-south trajectory (which would be desirable for technical reasons) is not socially realizable under the present circumstances.

Thus, this episode provides a striking illustration of the normative and political significance of the realization of closed technological systems. In MacKenzie's paper, however, the episode serves merely as an introductory example. It is not systematically analyzed and theoretically accounted for. The reason for this is that MacKenzie (1989, 412–418), following the sociology of scientific knowledge, understands the analogy between scientific experiment and technological testing primarily in terms of the notions of interpretative flexibility and closure of debate. He focuses on the theoretical interpretation and discussion of facts (about the artifacts) and thus tends to neglect the issues surrounding the (material and social) realization of closed (experimental and technological) systems.[10]

Next, I want to say something about Latour's point of view as it is propounded in his (1983) study of Pasteur. His view implies that a successful use of science in technological projects requires that the laboratory has to be rebuilt, as it were, as a whole on a larger scale in the open field. In that respect there would be no separation or distinction between inside and outside the laboratory. Moreover, Latour takes the scientist (in his case, Pasteur) to be the motor of the social transformations that are required for large-scale technological applications: "Pasteur actively modifies the society of his time . . ." (Latour, 1983, 156). This account, however, passes completely over the point argued earlier—namely, that closedness, and consequently success, cannot be measured in a context-independent way. Due to the differences between scientific and technological knowledge on the one hand and between the social conditions for experimentation and technology on the other, technological success is not identical with "scaled-up" experimental success. Moreover, the cases discussed in the present section clearly demonstrate that, in general, the realization of the social conditions required for the success of technological projects cannot possibly be accomplished by individual scientists.[11]

6.4 THE EFFECTS OF "EFFECT THINKING"

In the last section, my aim has been to support the proposed analyses with the help of examples taken from science and, especially, from technology. In this section, I want to explain their normative relevance with respect to public debates on technology and with respect to technology policy.

The conception of technology that prevails in these areas of public debate and policymaking is mostly framed in terms of ends, means, and effects. Theoretically, this conception has been supported by influential philosophers such as Popper and Habermas. The general contention of this section is that such conceptions of technology, to a large extent, fix and constrain what kind of reasons can count as "good reasons" in judging, accepting, or rejecting the realization of technologies. I will argue that in the light of the analysis of technology put forward in this chapter, the class of good reasons for evaluating the (possible) realization of technologies should be much wider than is the case in the dominant views.

Again, I will consider the example of nuclear energy technology, and I will focus in particular on the Dutch debate about it and on the view of technology that emerges from the reports of the BMD. The case is fairly representative for the following reasons. First, similar controversies also occurred in other countries and about other technologies (cf. Wynne, 1982; Nelkin, 1984); second, the notion of risk, which is a central element of the technology conception in the BMD reports, is also current in technology debates and technology policy elsewhere.

The Extensive Social Discussion, a combination of participation procedure and public opinion poll, took place between 1981 and 1983. It was led by a Steering Committee (operating independently of the government), and it consisted of two phases: an information phase, the results of which are laid down in an *Interim Report* (Stuurgroep, 1983), followed by a discussion phase, the results of which are incorporated in a *Final Report* (Stuurgroep, 1984). The intention of the BMD was to draw up a comprehensive and representative inventory of data, arguments, and preferences concerning possible energy policies.[12] Although officially the entire energy policy was down for discussion, the question of whether or not one should accept nuclear energy remained the pivot on which the debate hinged. Therefore I will restrict myself here mainly to that question.

The overall conclusion of the discussion about nuclear energy was clear enough: a great majority was against a future extension of the number of nuclear power plants and a sizeable part (varying between 26 and 58 percent, dependent upon the kind of poll or participation) moreover advocated the closing down of the existing plants in Borssele and Dodewaard (Stuurgroep, 1984, 188–195). As an alternative to nuclear energy, many people argued for a more extensive use of permanent energy sources, such as the sun, wind, geothermal energy, and so on (Stuurgroep, 1984, 89–91). However, these conclusions and the arguments upon which they rested were bluntly ignored by the then government and by a section of Parliament: in January 1986, plans for building two or more new nuclear power plants were being developed.[13] Clearly, the antinuclear energy movement had not been able to make capital of the course and outcome of the BMD: according to many, the most important (negative) result had been a depoliticization of the issue.

As I mentioned above, the focus of this section is on the conception of technology that can be extracted from the BMD reports. My criticism of it pertains, in particular, to the relation between cognitive and social aspects. For, appearances notwithstanding, in the *Interim* and *Final Report* these aspects have still been set apart. Yet this separation has been disguised rather subtly, so that it is small wonder that people complain of the depoliticizing character of the BMD without, however, "having been able to put their finger on the real blot" (Verhagen, 1984).

The main problem in this respect is that in both reports, social and environmental aspects of (energy) technology are analyzed merely in terms of *effects*. A typical sentence is: "The use of nuclear energy has a number of effects upon men and environment . . ." (Stuurgroep, 1983, 87; see also 122–128). Of course, it cannot be denied that technology has social as well as environmental consequences. For instance, the ejection of sulphur dioxide resulting from the combustion of coal has, through acid rain, definite effects upon the environment and (subsequently) upon human beings and society. However, an analysis in terms of effects only passes over the fact that successful technological production also requires the realization of material and social *conditions*. The production and maintenance of particular social conditions (for example, a bureaucratic and centralist administration in the case of nuclear energy) is necessary in order to be able to guarantee the permanent technological success of a project.

The difference between both conceptions is of fundamental importance. Effects, like the ejection of sulphur dioxide, can often be remedied by means of technological adjustments, without having disagreements about the relevant technology as such. Material and social conditions, however, are inherently connected with the endeavor for the production of closed systems; these conditions cannot be undone, without at the same time endangering the technological success. An important consequence of speaking only in terms of "effects," and not in terms of "conditions," is the reduction of sociopolitical issues to technical ones. The question "do we really want to fulfill the social conditions needed for this technology and the social control belonging to it?" is replaced by the question "with which adjustments can we keep the effect of this technology within certain bounds?" Indeed, this is a typical instance of depoliticization.

The way in which notions such as risk, risk analysis, and risk experience or risk perception are used in the BMD is likewise depoliticizing. In this context "effect thinking" again plays an important part.[14] "Risk" is, as usual, defined as a function (namely, the product) of "the probability of a certain, exceptional, event" and "the degree of undesirability of the effects of that event" (see Stuurgroep, 1983, 128). This definition implies that for arguments in terms of risk to make sense, one has to be able to calculate a *probability*. In the *Interim Report*, we find a significant series of judgments regarding this point. From: "the probability of a serious accident with nuclear energy cannot be estimated accurately" (Stuurgroep, 1983, 95); through: "the risk has to be described, estimated and assessed as well as possible" (Stuurgroep, 1983, 96); to: "the outcomes of risk analyses by experts give a very small probability of a very big accident" (Stuurgroep, 1983, 130). This series of statements is a striking example of what Hacking (1983a, 38, and 1986, 139–143) calls the transformation of "ignorance" into "(probabilistic) uncertainty." By doing so, a possible political decision (a choice based on social priorities) is transformed into a technical-scientific decision (based on rational risk analysis). In some passages of the *Final Report*, this line of reasoning is once more reinforced. If, on the ground of the present state of science and technology, unacceptable risks are virtually ruled out, the debate has to be closed: "further *uncertain* risks [as a consequence of ignorance] are to be accepted by everybody as a social burden."[15]

Another objection to the risk discussion in the BMD bears directly on effect thinking. After all, if we restrict ourselves to the notion of risk as it is described above, apparently only those events and those effects are undesirable that—from a technical point of view—are deviations from the normal functioning of the system of nuclear energy production. On account of the preceding considerations, this implies an inadmissible reduction of the problem: it should also be legitimate to open the discussion on the "risk" (for instance, the undesirable social conditions) of a technological system that is successfully operating in a technical sense.

In the BMD, however, a different strategy is pursued in view of the problems surrounding the notion of risk. It is argued that, alongside the risk analyses by experts, the "subjective risk experience of ordinary people" should play a part in decision making, in which, accordingly, a less formal and more qualitative notion of risk should be used (Stuurgroep, 1983, 130, and 1984, 252–255). Yet these additions (experience beside analysis, qualitative along with quantitative approaches) leave the framework of the discussions, the effect thinking, undisturbed and are, therefore, insufficient if one wants to reach a well-considered assessment of the realization of nuclear energy. Moreover, this strategy has an additional disadvantage, because it facilitates a division between, on the one hand, the experts, who are able to calculate risks more or less objectively, and, on the other hand, the general public, which only has a subjective, irrational awareness of risks.

Consequently, this strategy seems ill advised, certainly for those who are averse to, or critical of, the use of nuclear energy. For, once they have been moved over into the subjective and irrational corner, the field has been depoliticized, and they become an easy prey, within our "objective" and "rational" society. In the BMD, this move goes as follows. The Steering Committee recommends psychological research as a necessary supplement to quantitative risk analysis, but at the same time it states:

> Yet, psychological studies of the experience of risks have also raised doubts. Does one really say what one means during an interview? Is one's opinion not formed to a great extent by the information suggested by newspapers, radio and television? Is it not the case that there is an enormous lack of knowledge, with the result that one's opinion can hardly be reliable? (Stuurgroep, 1983, 130; see also Stuurgroep, 1984, 254)

These questions are unmistakably formulated in a rather tendentious manner. But our earlier analysis suggests that it is perfectly legitimate to pose exactly the same questions to the expert risk analysts and their instructors. Typically, this is not what happens in the reports. Apparently, the Steering Committee's intention with respect to the distinction between risk analysis and risk experience is different. For, in the *Final Report*, the (political) aim pops out of the (scientific) hat:

> more research will have to be done in the social scientific sphere as well, in order to examine how the big resistance against the introduction of nuclear energy may be obviated. (Stuurgroep, 1984, 212)

In conclusion, one can say that, certainly, in many cases individual fears and emotions will be a motive for resistance. In *social* debates, however, it is not these fears and emotions as such that should be focussed upon but rather the question of what are good reasons for acceptance or rejection of technologies and, accordingly, whether such reasons are present in the cases under discussion.

In this light, the results of this section can be summarized as follows. Which reasons count as "good" cannot be established in an a priori fashion; the matter depends upon the social context. One important element of that context is the technology conception underlying the debate or policy in question. In many discussions about technology, technological action and production are analyzed in terms of means, ends, and effects. First, the means are characterized as neutral, objective, and efficient (or inefficient); next, this characterization is sharply contrasted with the ends of the technological project, about which political and social debate is considered possible and necessary; while, lastly, the relationship between means and ends is evaluated in terms of the effects of the technological means at stake, which are, or are not, justified by the ends.

The above discussion implies a fundamental critique of accounts of technology that base themselves on such a conception. I have analyzed technology as (an attempt at) the production and maintenance of closed systems. Technology is only (can only be) instrumentally successful, if certain specific material *and* social conditions are fulfilled. Therefore, an assessment of technology not only concerns its effects but also and especially its conditions. In other words: instrumental action and instrumental reasoning can and may never be considered as nonsocial categories.[16] More specifically, this implies that the question of the acceptability of risks, defined in the standard manner, forms only part of the problem. Specific social "risks" are also involved, and it follows from the analysis of technology put forward in this chapter that a reference to these "risks" should count as just as much a "good reason" in assessing the acceptability of technologies.

6.5 THE INTRINSIC CONNECTION BETWEEN KNOWLEDGE AND POWER

Finally, I come to the question of the relationship between scientific or technological knowledge and social power. Work in the natural sciences, and in the technologies based upon them, not only consists of theoretical arguments but just as much of experimental interventions. True knowledge should, at least, be testable by means of the stable and reproducible realization of closed (experimental or technological) systems (cf. chapter 4). A successful realization of a closed system requires, however, not only scientific or technological knowledge and control of the relevant material conditions of closedness but also social knowledge and social control of those conditions. The social conditions are in general not fulfilled "by nature." In order to be able to fulfill them we need social power, and all the more so as the projects fill larger scales and longer terms. Put differently: one can only be sure that scientific or technological knowledge is invariably true if one assumes at the same time that the social power necessary for the actual realization of this knowledge in closed systems is or can be permanently exercised. In this way knowledge and power are *intrinsically* connected. That is to say: this connection is not a matter of an accidental or temporary limitation, but it concerns an essential aspect of natural science and technology, due to the requirements of experimental and technological stability and reproducibility.[17]

It follows from the analyses in this chapter that the power with which we deal here certainly does not originate from some sort of neutral knowledge (in the spirit of Bacon), neither exclusively from the scientists (as Latour would have it) nor from the professions of practitioners (as Illich claims). In general, this power will have its roots in society as a whole, and for every case we will have to examine what is contributed by, for example, the economy, the state, the professions, the experts, but also: the social movements, the protest groups, and even the students of science and technology. In this way we will be able to avoid both the Scylla of technological determinism and the Charybdis of social determinism. A central aim of the previous analyses has been to offer a theoretical approach that enables such a differentiated evaluation of this intrinsic power aspect of scientific and technological knowledge. Applied to concrete cases, we can first try to examine what kind of social control is required for the closedness of the relevant system, who exerts this control, whether one wants to comply with it, and how one can, possibly, resist it.[18] Further, this evaluation can be complemented by following the same procedure with respect to alternatives to the technology in question, in order to ascertain whether they are more acceptable or less acceptable.[19]

Let me close this chapter with three observations. First, the focus of my present account has been on the *connections* between power and knowledge. An explicit and more systematic analysis of the notion of power has not been aimed at. Yet, several aspects that should play a role in such an analysis emerge from the above discussions. For instance, the interdependence of the notions of power and control; the issue of the repressive and productive dimensions of the exercise of power; the institutionalization of power in scientific and technological organizations; the relationship between power and violence; the struggle between power and resistance and, hence, the limits of power; and the question of the normative legitimation of power. On the basis of the above analyses it is, I think, possible to deal with such crucial aspects of technoscientific power in a fruitful and differentiated manner.

Next, it will be clear that the purpose of the proposed approach is not primarily to provide a straightforward description of the actual historical development of experimental and technological practice or of the conscious intentions of actual scientists and technologists. Rather, the aim has been to offer some theoretical tools for analyzing and assessing the (successful *and* unsuccessful, conscious *and* unconscious) attempts at closing experimental or technological systems. To be sure, I did argue that closedness is a necessary condition for experimental or technological success.

But, whether or not an experiment or a technology is and will remain successful in this respect is a fully contingent matter.

Finally, as pointed out earlier, the notion of closedness bears upon a specific dimension of technological production. Consequently, an analysis and evaluation of technology as an attempt at the realization of closed systems does not exhaust the theoretical and normative issues that may and should be raised. For this reason a more general theoretical and normative approach is called for. In the next chapter, I will attempt to develop such an approach.

CHAPTER 7

The Appropriate Realization of Technology: The Case of Agricultural Biotechnology

7.1 INTRODUCTION

Normative issues in agricultural biotechnology are being widely discussed at present. Ethicists, philosophers, natural and social scientists, technologists, environmentalists, farmers' organizations, "third-world" groups, and governments contribute to the debate. These different groups approach the issues from their own perspective and on the basis of their own experience. Despite this, many of the discussions can be seen to be situated within a few common frameworks. Roughly, we can distinguish between an "ethical" and a "technical" framework for analyzing and resolving normative issues in biotechnology. Within the ethical framework, problems are mostly framed in terms of the moral acceptability of (the effects of) biotechnological products, whereas proposed solutions focus on individual actors and institutions and appeal to their (moral) responsibility. Within the "technical" framework, problems are usually conceptualized in terms of adverse side effects (for instance, hazards and risks) of the biotechnological processes, while solutions are sought in balancing such costs with the expected benefits.

In this chapter, we[1] argue that both the ethical and the technical frameworks are too narrow for adequately addressing the relevant issues. We do so by developing a more comprehensive account of what is normatively at stake in realizing technologies. On this basis we propose a notion of "appropriate technology" that is both theoretically adequate in analyzing ethical issues and practically applicable in assessing technologies and guiding our actions. To substantiate these claims, discussions of key features of modern technology will be combined with analyses of the problems and opportunities of appropriate technologies in specific cases of agricultural biotechnologies for small-scale farmers in developing countries.

We start our analysis, in section 7.2, by examining the gap between the claimed potentialities and the actual drawbacks of agricultural biotechnology, as well as some common ethical recommendations for bridging this gap. It is argued that the inadequacy of these recommendations stems from a deficient understanding of the development and use of technology. Therefore, section 7.3 explains what is required for, and at stake in, realizing technologies by means of a discussion of a number of "key features" of modern technology. On this basis we formulate, in section 7.4, our central questions as to what is normatively at stake in modern technology and what should therefore be on the agenda in assessing technologies in a just and democratic manner. In other words, these questions enable us to analyze (proposed) technologies in terms of their "appropriateness," and they imply, at the same time, suggestions for moving into a direction of greater appropriateness. In section 7.5, the relevance of the preceding account of appropriate technology is illustrated in detail by applying and specifying it for the case of agricultural biotechnologies for resource-poor farmers in developing countries. The final section sums up the main results and conclusions.

7.2 THE POTENTIALITIES AND ACTUALITIES OF AGRICULTURAL BIOTECHNOLOGY AND ITS ETHICS

Biotechnology involves the integrated use of molecular genetics, biochemistry, microbiology, and process technology employing microorganisms, parts of microorganisms, or cells and tissues of higher organisms to supply goods and services. This implies that biotechnology comprises a variety of different methods and techniques. These include fermentation technology, microbial inoculation of plants, biological control agents, plant cell and tissue culture, enzyme technology, embryo technology, protoplast fusion, hybridoma or monoclonal antibody technology, and recombinant DNA technology (see DGIS, 1989).

Many reviews claim that the potential applicability of biotechnology is virtually unlimited.[2] Agricultural biotechnology, for example, would be capable to address the needs of all categories of farmers, whether they are small scale or large scale, resource rich or resource poor. Farmers who can afford a high level of external inputs may be confronted with problems of ecological sustainability of their farming systems due to environmental pollution and with problems of economic sustainability due to uncertain prices of their outputs. In this case, so it is claimed, environmental damage can be reduced, for instance by developing new, pest-resistant varieties, while low and uncertain prices of outputs can be dealt with by producing new or improved biotechnological products. In contrast, the situation of poor and small-scale farmers in developing countries is quite different. For example, since they cannot afford costly external inputs, their farming systems are often threatened by soil depletion and even erosion. Here too, biotechnology claims to possess the potential for solving these problems, for instance by developing cheap and appropriate inputs that can help to maintain soil fertility. According to these reviews, this flexibility and adaptability also applies to other aspects of biotechnology. Both more traditional approaches, such as fermentation techniques, and more high-tech approaches, such as DNA recombinant technologies, are possible. Scientists and technologists from both private and public organizations can be involved in research and development activities. Production can be realized by large multinationals producing new seeds but also by family firms involved, for instance, in tissue culture. And biotechnology is easily adjustable to specific contexts, so that it can be used in all agroecological zones in the world. Thus, it is claimed that biotechnology has the potential to change agriculture, agricultural industry, and the world food production. It can provide practical solutions to a variety of problems in a large number of different contexts. In this way, it can make significant contributions to a balanced development, in which environmental, social, and economic issues can be incorporated.

In actual practice, though, agricultural biotechnology does not at all live up to these rosy reviews of its possibilities. As analyses of dominant trends show, without major changes, a balanced development is unlikely to occur. Instead, specific directions are explored while other possibilities are disregarded. In fact, agricultural products are mainly developed for resource-rich farmers with high purchasing power. Their wishes have been carefully mapped by means of marketing studies. However, the needs of small-scale farmers having few resources are hardly addressed or investigated, while their local knowledge and capacities are not taken into account. The main actors in the development of biotechnology are multinationals and certain public institutions, while governments are also heavily involved. In most industrialized countries (such as the United States, Japan, United Kingdom, France, Germany, and the Netherlands) developing biotechnology is given high priority. Governments endeavor to stimulate and influence biotechnology through grants to industry and public institutions. Companies, particularly large companies, have rapidly built ties with universities and public research institutes and employ them through contract research. Research contracts restrict the accessibility of public-sector research and provide an additional input for private-sector development of products. Several multinationals have attained a dominant position through contract research and the acquisition of biotechnological research firms. This position is protected by intellectual property rights (DGIS, 1991b). Thus, current biotechnological research and development programs are, by and large, guided by the economic considerations of a limited group of actors. Only those technologies are developed for which large and profitable markets exist. Examples are diagnostics, human pharmaceutical and animal vaccines, plant improvement techniques, and food processing procedures.

These trends entail several problematic effects. Three of them will be briefly mentioned here. The first is that the output of the large-scale farmers increases, while that of small-scale farmers remains virtually the same. The enhanced supply causes commodity prices to decrease, thus forcing small-scale farmers into an even more marginal position. Any excess production they might have had is worth less on the open market. Second, biotechnology has produced many novel, functionally equivalent substitutes for traditional raw materials (such as sweeteners for sugar). It has, moreover, enabled the production of existing raw materials in industrial rather than agricultural settings. Examples are the production of pyrethroids, vanillin, and the purple dye, shikonin, by means of plant cell culture. Practices like these involve further drawbacks for many farmers in the developing countries. Third, there is the issue of biodiversity. In general, the advantages of agricultural biotechnologies are considered greatest when they are applied in a standardized and large-scale manner. For this reason, biotechnology strongly promotes practices of monocropping. The direct and indirect consequence of these practices is a decrease in biological diversity.

The Inadequacy of Ethical Recommendations

The above discussion shows that in actual practice, agricultural biotechnology has a number of obvious negative impacts. This will not come as a surprise: the same or comparable conclusions have been drawn in many investigations before. The important issue then becomes: what can be done to reduce the negative impacts and to realize a more just development in agricultural biotechnology? In previous ethical reflections, particular recommendations have been made to address this issue (e.g., Jonas, 1979; Price, 1979). Further study, however, shows that the adequacy of these proposals is questionable. In general terms, the problem is this. The ethical recommendations bear on (parts of) the biotechnological processes. Yet, they tend to be unhelpful because they are not based on an adequate account of these processes. Instead, they are, mainly, confined to analyzing the (negative) effects of the biotechnological products.

To illustrate this, we will briefly examine some recommendations that are frequently offered to university scientists and industry (see Brouwer, Stokhof, and Bunders, 1992). One such recommendation to university scientists is not to engage in research that might contribute to the substitution of raw materials that are now produced in developing countries. Specific criteria for evaluating the acceptability of research with respect to the issue of substitution have been discussed in various universities. In these discussions several arguments against the implementation of such criteria were put forward:

- The academic freedom of scientists to study fundamental biological processes is threatened by such criteria.
- As the results of fundamental research are quite unpredictable, the applicability of the criteria is difficult or even impossible.
- Scientists are not responsible for the decisions of others to use the results of their work in a specific way.

The same recommendation has been proposed to industry. Again, a negative response resulted, although it was argued in different terms:

- The substitution of natural raw materials by industrially produced materials is not specific to biotechnology but a pervasive feature of all technological development.
- The problem of substitution is not caused by a single new technology, because it is rooted in a much broader economic and social context.
- Substitution does not always work in favor of the developed countries; for example, large amounts of vegetables and flowers, once grown in the Netherlands, are now produced in Egypt, Kenya, Morocco, and the Canary Islands (Veldhuyzen van Zanten, 1992).

Thus, this kind of recommendation focuses on what actors *should not do*. They should not be involved in research that might lead to substitution. As it turned out, the people and institutions addressed take this recommendation to be inadequate. They see their own involvement as only a marginal aspect of the whole substitution process.

Another kind of recommendation focuses on what actors *should do*. Here it is accepted that, generally speaking, biotechnology cannot be stopped, either by ethical calls to behave responsibly or in any other reasonable way. What might be influenced, though, is the way biotechnology is being developed and applied. Recommendations are made to strive for a more balanced development in which environmental and social concerns are taken into account next to economic concerns. University biotechnologists, for instance, are called upon to behave more responsibly and to orient their research not only to the wishes of industry but also to the needs of small-scale farmers.

In a further investigation of the implementability of the latter recommendations, university biotechnologists working with industry were asked to respond to the above analysis of trends and problems in biotechnology and to the call to work for small-scale farmers (Bunders, Sarink, and De Bruin, 1989). The outcome was that they generally agreed with the analysis but not with the recommendation. On their own, so they said, they were not able to focus their research on the farming systems of small-scale farmers and their specific needs and opportunities. They felt they did not have the required expertise and lacked possibilities for obtaining the necessary information. For example, a professor of microbiology stated that his field (nitrogen fixation) could be useful to small-scale farmers in developing countries and that he would react positively to research questions that might contribute to the needs of the rural poor. The initiative for this, however, should be taken by the developing countries, because he himself found it hard to set up a project in such a way that it had a chance of being useful to the small-scale farmer.

The results of these interviews are supported by a larger study (De Bruin and Bunders, 1987). Thirty-five plant biotechnologists working in Dutch governmental institutions and universities were interviewed and asked to fill in a questionnaire. Most biotechnologists (70 percent) had never had any professional contact with development, environmental, or small-scale farmers' organizations. Almost all of them (90 percent), however, maintained contacts with industry. Even so, most interviewees felt that biotechnological research could be adjusted to the needs of groups other than industry, such as farmers in developing countries. According to them, the main reason why this does not happen on any significant scale is the lack of financial support, the lack of the necessary knowledge, and the lack of contacts with relevant organizations.

Thus, quite generally, the actors addressed feel that they are not able to act according to the ethical recommendations.[3] Most of them really do not have any idea of how the dominant trends in biotechnology, which make the rich richer and the poor poorer, might be influenced. The problem with these recommendations is that they are too general and do not specify what actors can and should do in practice. In other words, these ethical evaluations state, in abstract terms, *what* (individual) scientists ought to do, but they fail to make explicit *how* this could be done. The same problem has been noticed by Crocker in an extensive review of "development ethics." According to him, "ethicists must remember that to know what ought to be, we must first know what can be. For, as Kant said, 'ought' implies 'can'" (Crocker, 1991, 467). Consequently, he criticizes the restriction of moral judgments "to one's private closet or to an inflexible professional code" (Crocker, 1991, 463), and he argues that moral reflection should never be "isolated from the sciences, politics and practice of development" (Crocker, 1991, 467).

In the same vein, the problem with the above-mentioned ethical recommendations is that they are not based on an adequate account of what is required for, and at stake in, the whole process of realizing agricultural biotechnologies. Instead, they appear to assume that, given the claims of the almost unlimited potential of biotechnology, the realization of this potential is a matter of straightforward "application" of available knowledge and techniques and thus it requires little more than the willingness of the individual scientists or institutions that are involved in the process. Such a view, however, greatly underestimates the quantity and quality of the work that is generally needed to realize the potentialities of nature in technological projects. Thus, if we want to propose adequate ethical recommendations for a more appropriate agricultural biotechnology, we need to make a detour through the complexities of modern technological systems. Doing this, however, implies that we cannot stay within the boundaries of common ethical approaches, since it requires an integration of knowledge and expertise in the fields of biotechnology, technology assessment, science and technology studies, and ethics.

7.3 REALIZING TECHNOLOGY

What is required for, and at stake in, realizing technologies? A satisfactory answer to this question presupposes a more or less articulated account of what technology is. What is clear, though, is that technology is a complex and multifaceted phenomenon that may be studied from many different perspectives. Accordingly, many distinct disciplinary approaches to technology have been taken. Thus, history, philosophy, sociology, economics, ethics, and policy studies have all examined technology and revealed various important dimensions of it. Yet, notwithstanding the existence of such diverse approaches, technology possesses a number of significant characteristics that should be dealt with in any adequate account. Therefore, these "key features" are the starting point for our discussion of appropriate technology.[4]

A *first* key feature is already hinted at in the title of the present section: technologies have to be *realized*. Although this may seem evident, it is well worth noticing. It says that technology is neither a naturally occurring process nor a process that takes place in the realm of ideas or in the mind of people. Realizing a certain technology requires concrete, active, and often laborious interventions in *particular* parts of the world. This suggests that conceptualizing technology in terms of a "given" potential and an unspecified but virtually unlimited set of "applications" tends to systematically underexpose the significance of the realization processes (cf. chapter 4). However, in studying technology, the questions of "where," "when," "by whom," and "for whom" a technology is realized are of utmost importance. Furthermore, the notion of realization also implies that it takes time to realize a technology. Again, this seemingly obvious fact is not without significance. The shorter or longer period needed to realize technologies imparts them with a certain "endurance": a technology, once realized, fixes the nature of the relevant part of the world in a specific way. This fact sets clear boundaries to the exercise of freedom and democracy, especially in the case of large-scale technologies. When we have realized an extended road system for massive car traffic in a densely populated country such as the Netherlands, choosing a qualitatively different way of transport and implementing it within a reasonable time span will turn out to be hard indeed. Or consider this example. In Zimbabwe, at the time of its independence in 1980, the most important agricultural research institutes were located in large-scale commercial farming areas. This was a direct consequence of its colonial past. The time required for reorientation of these institutes towards the needs of small-scale farmers in the communal areas has been estimated to be at least ten to twenty years (cf. Bunders, 1990, 129 and 158). In sum, particular technologies embody particular and relatively enduring commitments to specific ways of life.

Second, technological realizations possess a *systemic* character. That is to say, technologies should not be seen as separate, isolated things or artifacts. Instead, any adequate understanding should take account of the issue of the systemic interaction between, and coordination or integration of, different components into technological systems. Of course, in practice, systemic coordination or integration cannot be taken as given or unproblematic. Rather, it presents, for those who wish to have the technology realized, a task to be accomplished without there being any guarantee of success. Both small-scale and large-scale technologies possess a systemic character. For instance, if the goal of a mechanical typewriter is taken to be the production of clearly readable texts in a reliable and relatively efficient manner, its successful realization (which includes manufacture, maintenance, instruction, and use) requires the systemic interaction, coordination, and integration of a number of material and social-organizational processes. In the case of large-scale technologies, the systemic character is still more evident. Consider, for instance, the electricity supply system (which is itself a component of nearly all modern, large technological systems). The realization of an electric power system requires the creation and maintenance of an extensive and integrated sociotechnical network for the generation, transformation, transmission, distribution, utilization, and control of its product, electricity.

Phrases such as "material and social-organizational processes" already point to a *third* key feature of technology. They show that the components of technologically realized systems are not just "different" but *heterogeneous*. In examining technological systems we meet with physical artifacts (the keys of the typewriter; the electricity conducting wires); organizational arrangements (mutual attuning between sales and uses of typewriters and production of correction fluids; public or private organization of the generation of electricity); individual people (buying a typewriter for home use; applying for a job in the financial department of an electricity distribution firm); government institutions (legal approval for schooling professional typists; setting standards for the safety of electrical switches); knowledge (the know-how of the typewriter repairer; scientific calculations about the heat loss during the transmission of electricity); normative recommendations ("do not type application letters"; "be aware of environmental issues in purchasing and using electrical devices"). Clearly, the heterogeneity of technological systems—with components ranging from physical artifacts to normative recommendations—illustrates their complexity. The above examples also show that the notion of "system" is used here in an unassuming sense, primarily intended to highlight the issues of interaction, coordination, and integration. Moreover, stressing the features of realization and heterogeneity counteracts the tendency—which is often present in system accounts—to ascribe an autonomous teleology to technological systems and to deal with success and failure in a purely "technical" manner.[5]

Looking specifically at the development of *modern* technology, the local realizations of heterogeneous technological systems display a number of significant, nonlocal patterns (cf. section 5.4). For the purposes of this chapter, three further key features that are implied by the existence of such patterns will be considered. A *fourth* key feature of technology is the increasingly important role played by *science*. This includes not just explicit scientific knowledge, expressed in empirical or theoretical statements, but also scientific know-how, embodied in theoretical, experimental, or observational skills and techniques, and materialized scientific results, such as specific substances, materials, or instruments. The fact that modern technology and modern science are so thoroughly interwoven necessitates substantive processes of "translation" between the scientific and the technological context (cf. Latour, 1983, 1987b; chapter 6 here; Rouse, 1987, esp. 226–234). In general, such translations will involve a two-way interaction between both parts of the technological system and not a mere one-way "application" of science to technology. Thus, scientific knowledge, skills, techniques, and materials do not, by themselves, "diffuse" from the laboratory to the factory and the user. On the contrary, much hard work is needed to transform a successfully realized scientific result and to incorporate it into a working technological system. For this reason, scientific success does not automatically imply technological success. Typically, the larger the differences between scientific setting and technological system, the harder it will be to successfully translate the scientific results to the technological context. Thus, this key feature of modern technology once more underlines that the issue of where to realize a technological system is a crucial one. There is, for instance, a world of difference between realizations of "the same" technology in a poor developing country or a rich Western country.

A closely related, *fifth* key feature is *division of labor*. This phenomenon is especially apparent in large, science-based technological systems. A science-based technology implies a considerable division of labor among those who develop, produce, operate, and use it. The specific knowledge, skills, techniques, and materials required for realizing or maintaining science-based technologies are by no means generally available. This implies, for the large majority of the users of these technologies, a relationship of *dependence* upon other's expertise and facilities. When the price of correction fluid is suddenly doubled or its supply stopped, the individual user of even a simple typewriter feels this dependency. At the level of countries, larger effects of this division of labor will be felt, for example when costly facilities and expertise have to be imported and paid for during prolonged periods. Certainly, some kinds of division of labor and resulting dependencies have always been present in all societies. However, both the degree of specialization in modern science and the worldwide extension of science-based technologies have dramatically increased the significance and scale of these technologcal dependencies. As a consequence of the process of division of labor and in response to the problems met in attempted translations, science-based technologies are more and more standardized and increasingly display a black-box character. Many technological artifacts are locked up in "black boxes" so as to be stable against a number of variations occurring in their different contexts of use. And many components of technological systems are standardized, in order to be utilizable by relatively unskilled users. Thus, on the one hand, for most people involved in modern technological systems, their (inner) working is to a large extent invisible and inscrutable. This is what Borgmann (1984) has called the "device character" of modern technology (see also section 2.7). On the other hand, the notion of realization implies that no technology can work unconditionally. Therefore, its use will always require some knowledge and skill. Handling domestic electrical devices, for instance, requires the ability to understand and realize the installation and operation instructions.

A *sixth* important tendency of modern technology is its *expansionism*. This has to be taken in two senses. First, there is the fact that modern technology has become heavily involved in an ever increasing number of areas of ordinary life. Food production and consumption, communication, the conduct of war, making music, human reproduction—in all these areas technology has had a substantial impact and has brought about radical changes. Therefore, the organicist conception of technology as no more than a means for the amplification or replacement of the powers of the human body has turned out to be much too restrictive. The same holds good for the (broader) view of technology as a means for the material survival and reproduction of the human species. Second, technology is expansive in that it spreads over a larger and larger part of the earth. Electricity and telecommunications systems reach out to ever more remote areas. The same transistor radio can be bought and used in Los Angeles, Swerdlowsk, and Harare. Because of the expansionism of modern technology, in both of its senses, some authors (e.g., Borgmann, 1984) even claim that we live in a "technological culture."

7.4 REALIZING APPROPRIATE TECHNOLOGY

From the preceding section, we conclude that modern technology can be characterized as an attempt to realize heterogeneous technological systems, which prove to be increasingly science based, dependent on division of labor, and expansive. The first three key features (realization, systemic character, and heterogeneity) can plausibly be considered as characteristic of all technology. In contrast, the last three (expansionism, dependence on science, and division of labor), which appear to be typical of modern technology, are a result of contingent historical developments. The next task is to deal with the normative issues that are at stake in realizing technologies. More in particular, the aim is to develop an adequate and practically applicable interpretation of the notion of "appropriate technology."

The first question concerns what is normatively at stake in attempts at realizing technologies. Consider a particular technological system that has been, or might possibly be, realized at some spatiotemporal location. The notion of realization implies that we should take into account not merely the products but also the entire process by which these products have been (or might be) realized. Then, given the fact that realization implies an active intervention at particular spatiotemporal locations, a successful technological realization of the system requires a specific control of the relevant parts of the world. And, given the fact of the systemic heterogeneity of technologies, this control not only pertains to the material reality but equally to the psychological, social, and cultural reality of the actors involved. With the help of the first three key features, these major normative questions can be formulated:

1a. Are the products resulting from the realization of the technological system desirable?
1b. Are the new choices and possible courses of action resulting from the realization of the technological products desirable?

2. Are the (material, psychological, social, and cultural) conditions, required for the successful realization of the technological system, feasible and desirable?
3. Is enough known about the realization of the technological system to even pose the preceding questions in a sensible way?

As will be seen in more detail in the next section, the key features of science dependence, division of labor, and expansionism disclose a range of significant normative or normatively relevant issues, which should be taken into account in specifying and answering the above questions.

Questions 1a and 1b are related to each other. Both bear on the products provided by the technological system and focus on their desirability, independently from the particular way they have been realized. The products in question may be both intermediate and ultimate products of the technological system. Question 1a asks for the desirability of the products "as such." It leads to a number of important, normatively relevant issues. For instance: Who desires a particular product: producers, individual users, social organizations, government institutions? Who assesses the quality of the product: multinational companies, governmental regulatory bodies, consumer organizations, the general public? Which criteria count as criteria of quality and significance of the product: durability, "glossiness," competitiveness (in various senses of the word), environmental sustainability, cultural appropriateness? Further normative aspects of the realization of technological products are dealt with by question 1b. This question concerns the normative issues with which users are confronted as a consequence of the availability of the product. Generally speaking, a new product will enable new courses of action and therefore requires new kinds of choices. Should one, in order to discuss a somewhat delicate matter with somebody, approach him or her personally, or can it be dealt with by telephone? Does the speed of air transportation outweigh the pleasure of other means of traveling? Do we really wish to be able to choose in advance the sex of our children?

Question 2 is a complex question that pertains to the realization of the entire technological system in its (intended) spatiotemporal location. It concerns the normative or normatively relevant conditions that have to be fulfilled for the successful realization of the technological process. Generally stated, the normative question is whether or not the (results of the) interventions in material, psychological, social, and cultural reality—which are needed for the successful realization of the technology in question—are and will remain feasible according to the best of our knowledge and desirable for the people involved in the process. The contrast with questions 1a and 1b will be clear: the present question may also focus on the desirability of the products, but even so the relevant issues will always be dealt with in the context of the entire realization process.

In relation to this question, we can distinguish between two groups of issues (cf. chapter 6). A first group concerns what is normatively at stake in case of technical success. Here are some pertinent questions that fall under this heading: Can the employees of a nuclear power company be permanently motivated or disciplined to accurately carry out all the tasks required for a controlled operation of the entire energy production cycle? Is the continuing dependence on Western knowledge and skill, required for the successful tranfer of a complex technological system to a developing country, socially and politically acceptable? Will it be possible to maintain a constant product quality, despite the physical, psychological, and social "disorders," caused by the irregular work in a day and night shift system (which is, in turn, necessitated by a particular chemical production process)?

A further group of normative issues concerns the possible effects of the attempted realization of the technology, not in case of success but in case of failure. Since no technological system works perfectly, and many work quite imperfectly, failures, breakdowns or accidents may be expected frequently. Here too, one is confronted with a number of weighty and normatively relevant questions. What would happen if a large-scale accident occurs in a particular type of technological system? Are some types of technological systems more "accident-prone" than others? Why are nuclear power plants so often situated close to the borders of a country, or: *who* runs the risk in case of technological disasters? What does it mean when risk analysts claim that ordinary people's perceptions of the possible hazards of new technologies are often "irrational"? How are regulatory safety norms established, applied, and enforced?

These more specific questions also show that feasibility and desirability may variously relate to each other. They can be either independent or dependent. On the one hand, the realization of a technology at a particular spatiotemporal location may be feasible but not desirable, as in the case of an enforced realization of the relevant social conditions; conversely, however desirable a technology may be for all the people involved, it will not be feasible when the relevant material conditions turn out to be systematically unrealizable. On the other hand, when the force applied for the purpose of realizing the relevant social conditions is successfully resisted, a technology will be not feasible because it is not desirable; and finally, when a technology is confronted with unacceptable technical failures, its realization will be undesirable because it is not feasible.

The third question is, in fact, a preliminary question. It should be answered first in order to see whether it is at all sensible to *start* looking for answers to the other questions. Do we have at our disposal sufficient and reliable information and knowledge for a reasonable evaluation of proposed answers to the other questions? In dealing with this question, it may be helpful to distinguish between a straightforward, de facto lack of information and knowledge and the kinds of ignorance that stem from more fundamental fortuities. This distinction is quite relevant with respect to the issue of what to do when confronted with a lack of knowledge. In the former case, we may attempt to solve the problem by a (closer) examination of the subject in question. The latter case, however, provides strong reasons to forbear from realizing the technology in question, especially when its possible impacts can be expected to be large. Thus, Hacking (1986) argues that realizing a technology that exhibits significant "interference effects" (that is, effects that are essentially uncontrollable) makes the people involved liable on the basis of a "culpable" ignorance. Also in discussions about the release of genetically modified organisms in the environment, a distinction between de facto and "more fundamental" ignorance appears to be pertinent.

The above questions, together with the key features discussed in the preceding section, constitute a framework for dealing with normative issues in technology that is wider than the "ethical" and "technical" frameworks mentioned briefly in the introductory section. Roughly speaking, traditional ethical approaches focus mostly on questions 1a or 1b, while many studies in the fields of risk analysis and technology assessment merely address the second group of issues mentioned under question 2. In contrast, we want to emphasize that normative issues regarding technology do not merely pertain to "moral choices," "adverse side effects," or "costs and benefits." What is at least as important in a normative evaluation of (proposed) technologies is the *quality* of the natural, personal, and sociocultural world in which the people involved will have to live in order to successfully realize the technologies in question. The point also has implications for the meaning and scope of question 3. After all, when a technology is conceptualized exclusively on the basis of the ethical or technical framework, almost inevitably there is a lack of knowledge regarding aspects of the technological system that are relevant to answering a whole range of normative questions. Apart from this, the proposed framework entails another advantage over more common ethical approaches. As we have seen in section 7.2, many of the usual ethical discussions and recommendations turn out to be unworkable. An important reason for this is their exclusive focus on the decisions of isolated individuals or institutions. In contrast, the above analyses imply that a more adequate approach should primarily take account of the essential but precarious interaction, coordination, or integration between the heterogeneous components of technological systems (cf. also De Vries, 1991).

Appropriate Technology Defined

We can now sum up the discussion by defining a technology as *appropriate* if the three questions stated above can be answered in the affirmative by all of the people involved in the realization of the technological system.[6] We are aware of the fact that in actual practice it will be difficult to realize technologies that are fully appropriate in the above sense. Therefore, our notion of appropriateness should be seen as primarily regulative. It bears upon far-reaching and highly significant issues of our age. And it makes explicit what is normatively at stake in modern technology and what should therefore be on the agenda in assessing the feasibility and desirability of technological systems in a just and democratic manner. Such assessments can be made both in evaluating the appropriateness of existing or earlier technologies and in defining the issues that should be taken into account in the (re)search for and development of appropriate new technologies. In many views it is, implicitly or explicitly, assumed that "appropriate technology" is primarily an issue for developing countries. The presupposition of these views is that technologies in developed countries are, more or less by definition, appropriate. In contrast and in spite of the particular case study presented in the next section, the notion of appropriate technology proposed here is meant to be generally applicable.

Of course, in attempts at actually realizing appropriate technologies, we will always be confronted with established interests and asymmetrical power relationships, as well as irreducible disagreements about what is feasible and desirable. Consequently, in practice "appropriateness" will be a matter of degree. For this reason, our normative "recommendation" to the actors involved in the realization of technological systems is to attempt to increase the measure of appropriateness by trying to decrease the degree of inappropriateness. As will be seen in the next section, this will mostly amount to pinning down and criticizing less appropriate aspects of technologies and clarifying the conditions under which more appropriate realizations might be obtained. On the present analyses, more appropriate technologies can be realized in two different ways: by giving more of the people involved a (greater) voice in finding out what is feasible and desirable; and by realizing technologies that involve fewer people (or involve the same people less). The former approach seeks to alter some of the asymmetrical power relationships, whereas the latter tries to diminish the effects of the genuine disagreements with respect to feasibility and desirability. A further question, then, is how to organize such approaches. So far, several organizational models have been proposed and tried out (see, e.g., Goggin, 1986; Bunders, Stolp, and Broerse, 1991; Laird, 1993). An important task in trying to realize more appropriate technologies concerns the theoretical development and the practical implementation of such models.

Viewed in this way, the above notion of appropriate technology provides at once a valuable ideal and a practically helpful framework in assessing technologies and guiding our actions. Our account involves several normative and normatively relevant premises. The first is procedural. Given the large impacts technology has on the worlds in which people live, all people ought to have a voice in the way their world will be shaped by the realization of new technological systems. Moreover, answering the three questions will mostly require an active participation of the people involved rather than a simple registration of their wishes. Indeed, the ought-implies-can principle entails that establishing the appropriateness of a technology requires an engaged investigation of its feasibility in the (intended) realization contexts and not just an abstract examination of its desirability. This aspect of our approach allows for the fact that interests and desires are not immutable givens but may be transformed in the participation process to a higher or lesser degree. In the second place, our account does not presuppose a separation between moral and other normative issues. The concept of realization of heterogeneous systems implies that moral issues will always be intrinsically connected to other normative questions, be they cognitive, political, sociocultural, aesthetic, or otherwise. Therefore, the possible resolutions of particular issues will also be heterogeneous. Third, we suggest that any normative approach should not only be judged on what it explicitly says but also on what it implicitly shows. In this sense our focus on the problems and opportunities of the poor and small-scale farmers in developing countries expresses a further moral premise. In the next section, these normative and normatively relevant claims will be illustrated in detail in a discussion of cases of agricultural biotechnologies.

Finally, it should also be clear what the notion of appropriate technology does *not* imply. For one thing, it does not presuppose a simple dichotomy that would enable us to distinguish once and for all between "intrinsically good" and "intrinsically bad" technologies (cf. Winner, 1986). Consider, for example, the list of "characteristics of appropriate technology" proposed by Barbour. Appropriate technology would always be of intermediate scale, labor intensive, relatively simple, environmentally compatible, and locally controllable (Barbour, 1980, 296–298). In our view, such global characterizations are less than adequate. Thus, Barbour's claim that labor intensive technologies are the right ones for developing countries—since labor would be plentiful in the third world—is not generally true. To mention just one example, it does not apply to Zimbabwe, where "lack of labour is one of the major constraints in the small-scale agricultural production" (Bunders, 1990, 160). Moreover, appropriate technology can no more be defined by an exhaustive list of artifacts than by an unambiguous list of characteristics. For example, consider the wood stove, an item frequently advertised by the appropriate technology movement in the United States during the 1970s and early 1980s (see Winner, 1986, 70–80). In practice, many of these stoves are fed with industrially processed wood that contains various poisonous substances. Consequently, large-scale use of this "appropriate" technology would result in large-scale pollution. Thus, this example once more illustrates an important point made above. In evaluating the appropriateness of technologies, one should not be restricted to isolated things but explicitly take into account their systemic connections.

7.5 REALIZING APPROPRIATE AGRICULTURAL BIOTECHNOLOGY

The general analysis of appropriate technology will now be applied to the case of (intended) realizations of agricultural biotechnologies, especially in developing countries. As concluded in the preceding sections, an adequate treatment of the relevant normative issues considerably transcends what is usually considered as the proper field of ethics. Moreover, our goal is not merely to evaluate the technologies in question as to their acceptability but also, or even primarily, to offer concrete clues on how to proceed towards greater appropriateness. Since it is of course impossible to deal with each and every aspect of our topic here, the discussion is restricted to specifying and illustrating some of the most important issues. The approach will be to discuss the normative questions posed in the preceding section and to examine a number of significant normative problems in terms of the key features of technologies as described in section 7.3. In doing so, the focus will be on opportunities and problems of the small-scale and poorer farmers in developing countries. More particularly, points made will be illustrated by reference to the case of a yam tissue culture project set up in the Caribbean in the early 1970s (Bunders and Broerse, 1991; see the box for a short description of this project).

To facilitate discussion, question 3 of whether sufficient knowledge is available for the realization of a technology will not be dealt with separately. As pointed out earlier, this question is actually a preliminary question that is relevant as soon as one begins thinking about the other questions. Therefore the issue can be more conveniently discussed in the context of the other questions (1 and 2). In the yam tissue culture project, for instance, cases will be encountered where lack of knowledge and information made the project inappropriate from the start. More generally, in thinking about intended biotechnological realizations, especially when they are complex, large scale, and long term, we should be aware of the issue of ignorance. Who knows what, in the long run, will happen to genetically modified organisms released in the environment? The explicit inclusion of our third question is meant to remind us that in cases where a lot is at stake about which we are basically ignorant, not realizing the intended technologies might well be the most desirable option. Let us now turn to a discussion of our first question.

Yam tissue culture project in the Caribbean
(based on: Bunders and Broerse, 1991, 26)

During the 1960s, the increasing incidence of internal brown spot and several other diseases was having serious effects on yields and quality of yam tubers in Barbados and other countries in the region. This created shortfalls in yam production for indigenous food and problems for the growing export trade. Yam is an important subsistence crop in the Caribbean. It is grown mainly by small-scale farmers on plots of less than 0.5 hectares, and it contributes as much as 50 percent of the dietary calories in the region. The crop is also an important source of export revenue and a feedstock source for local industrial processing.

In 1973, in response to the disease problems, a project using the new techniques of tissue culture started. Tissue culture was to be used to try to eradicate the disease and to multiply virus-free seed tubers for distribution to local farmers. The project initially had the backing of the U.K. Overseas Development Administration. Its early results were very promising, and, with the support of other donors, the project's objectives expanded to include the establishment of a self-financing yam tissue culture and propagation laboratory (the "Tissue Culture Unit") in Barbados under the aegis of the "Caribbean Agriculture Research and Development Institute." By 1982, disease-free clones of the popular species *Dioscorea alata* were available to farmers through an "approved growers" scheme. Average yield gains at farm level were about 30 to 40 percent, and the yam tubers were of consistently high quality. Yield gains as high as 95 percent were reported from initial trials in low-technology small-scale farmer production in St. Lucia.

After the project ended in 1984, however, things started to go badly wrong. Approved growers took to multiplying the improved varieties conventionally in the fields rather than returning to the Tissue Culture Unit for new disease-free stock. This, together with a number of other factors, may have contributed to a serious outbreak of the yam foliage disease "anthracnose" during the mid-1980s. The yam variety originally chosen for the tissue culture work turned out to be especially susceptible to anthracnose. At the same time, the viral disease, which the original project had been designed to eliminate, reappeared. Because of these difficulties, the Tissue Culture Unit never achieved its objective of self-funding.

Question 1a: Are the products resulting from the realization of the technological system desirable? And question 1b: Are the new choices and possible courses of action resulting from the realization of the technological products desirable?

It is instructive to first examine these questions for the case of the Caribbean yam tissue culture project. As explained in section 7.4, the above questions regard the (intermediate or ultimate) products provided by the technology, and focus on their desirability, independently of the particularities of the realization process. In the tissue culture project, the intended products were disease-free yam seed tubers to be used by local farmers in the Caribbean. These products turned out to be desirable to many small-scale farmers, for whom the diseased yam tubers did constitute a genuine and significant problem. Furthermore, the variety selected for initial research, *Dioscorea alata* c.v. White Lisbon, was grown by small-scale farmers. In fact, however, this "appropriateness" was to a certain extent a matter of good luck, since the small-scale farmers were never explicitly identified as the target end users.

Yet, the desirability of products can also be looked at from a different perspective. Consider the following questions. Is it desirable to aim at developing appropriate technologies for the mass of poor farmers living in "low potential areas," whereas a small number of other farmers possess large farms in the "high potential areas"? Do such projects, in other words, not simply legitimize the unjust division of land that has often resulted from Western colonialism? And would it not be better to support projects that aim at a more balanced division of land instead? These questions have in fact been posed at the start of a project for small-scale farmers in Zimbabwe. Generally, the farmers' response was that they felt that they had to strive for both options. Thus, they did not consider attempts at realizing biologically improved agricultural technologies as being a priori undesirable.

Unfortunately, not unfrequently intended technological realizations are inappropriate from the start. For instance, this is the case when technologists set out to provide a product that solves a "problem" that is not considered a problem at all by the prospective users. A telling example can be found in the area of fermented foods. Fermentation of food is important for small-scale farmers, because it increases the value of raw materials and extends their shelf life. Thus, university scientists attempted to discover ways to improve the fermentation process of soya (Prasetyo, Timotius, Stouthamer, and Van Verseveld, 1991). After the research had been completed, however, it became apparent that small producers had different quality criteria from those that had been assumed. *Taste* turned out to be crucial, instead of nutritional value and digestibility (the quality criteria of the World Health Organization). Comparable examples can be found in the field of plant breeding. Cassava comes in two varieties: bitter (toxic) and sweet (nontoxic). Because of this, biotechnological projects have been proposed to modify the toxic bitter cassava into the nontoxic sweet variety (Bunders, Broerse, and Stolp, 1989). The latter product, however, is not seen as desirable by the prospective users. According to them, the bitter cassava's toxic nature can be dealt with by further processing, while this disadvantage is outweighed by the fact that it has a much longer shelf life (DGIS, 1991a).

Such a lack of focus on the needs and wishes of the poorer farmers is a recurrent phenomenon in biotechnological projects. It is difficult to find biotechnological projects initiated by and directed at resource-poor farmers. A necessary condition for the appropriateness of any technology would seem to be to establish the desirability of the intended product in a substantive dialogue between its prospective users and the other actors involved. Engaging in such a process assumes that the needs and wishes of the users will be taken into account from the start. Given the great "distance" between, for example, Western laboratory life and the life of small-scale farmers in developing countries, setting up such a dialogue is not an easy task. Moreover, it implies that the initial views, plans, and wishes of the participants may be modified in the course of the process. The scientists may become convinced that their plans simply do not fit the needs of the farmers; and conversely, these farmers may develop new perspectives on their situation and find their needs and wishes transformed.

In concluding the discussion of the first question, let us briefly mention some aspects of its second part, which regard the desirability of new choices and possible courses of action resulting from the realization of agricultural biotechnology. Typically, biotechnology has and will have a large variety of impacts. Its most far-reaching effects are well known: product modification, product substitution, overproduction, and loss of genetic diversity. Clearly, these effects (will) lead to novel choices and new courses of action that are by no means generally desirable. This does not mean, however, that biotechnology cannot contribute at all to new courses of action that are considered to be desirable. For example, new or improved fermentation technologies that increase the storability of raw materials may be important for farmers in developing countries, who cannot easily transport their products to the market. Again, however, a substantive interaction between the producers and the users of such an intended technological realization is a necessary condition for the appropriateness of its product.

Question 2: Are the (material, psychological, social, and cultural) conditions, required for the successful realization of the technological system, feasible and desirable?

This important question pertains to what is normatively at stake in the entire process of realizing heterogeneous, technological systems.[7] As such, it deals with wide-ranging and complex issues. In many studies, especially in the fields of risk analysis and technology assessment, these issues are being discussed exclusively or primarily in terms of the (acceptability of the) hazards and risks resulting from the (intended) realization of the technological systems. From our point of view, however, to deal with the above question merely in these technical terms implies an inadmissable reduction of the issues that should be taken into account in assessing the appropriateness of technologies. First, the occurrence and the deeper causes of technical failure and risk cannot be adequately understood in isolation from its social and cultural situation. Second, there may be weighty reasons for judging a technology inappropriate, even if its realization does not entail any significant hazards or risks. Therefore, as demonstrated below, it is more adequate and more fruitful to discuss the issue of hazards and risks within the context of the more comprehensive question 2.

The approach adopted in dealing with this question will be as follows. The realization of agricultural biotechnological systems comprises a large number of heterogeneous components: relevant biological knowledge, skills, techniques, and materials; farming systems in their social circumstances; culturally mediated normative premises; individual people (farmers, researchers, policymakers, environmentalists) with their various specific needs and wishes; governments, governmental and nongovernmental organizations in developed and developing countries; national and international industries having particular interests and strategies; markets and price policies with their problems and opportunities; etc. (see, e.g., Bunders, 1990). Obviously, it is not possible to deal with all these aspects within the confines of the present chapter. Therefore, a number of particular issues will be selected for further discussion. These issues will then be analyzed with the help of our key features in order to examine their problems and investigate their degree of appropriateness. Again, explicit attention will be paid to difficulties and opportunities of changes towards greater appropriateness. Along these lines, we will subsequently deal with farming systems in developing countries; the input production systems for such farmers; the dependence on science; the commercialization of public biotechnological research; and, finally, the issue of how to organize appropriate systemic interactions. In all cases our prime example will, once more, be the yam tissue culture project.

Farming Systems in Developing Countries

Intended innovations in agricultural biotechnology are claimed to be potentially appropriate for resource-poor farmers because they would be inexpensive, uncomplicated, and would not require capital- and energy-intensive inputs. Although these claims may be justified in some cases, it is impossible to predict in advance which of the range of biotechnological developments might benefit the small-scale farmers without looking in detail at their local situation. Are the conditions required for the successful realization of the technology at the farming systems feasible and desirable? In examining this question one should be aware of the fact that farming systems also comprise a number of heterogeneous and interacting elements. Involved are biotic elements (plants, animals, microorganisms), abiotic elements (water, temperature, soil texture, nutrients), other "internal" elements (draught power, land, processing facilities, agricultural knowledge and practices, availability of labor, social relations, cultural belief systems) and "external" elements (fertilizers, seeds, market changes, institutional linkages, socioeconomic and cultural relations with other regions).

Again, the yam project clearly illustrates how well-intended technological projects may become inappropriate when the specific farming situation of a large group of prospective users is not explicitly taken into account right from the start. Consider the following chain of events. Initially the project was quite successful in terms of yield gains and improved quality of the yam tubers. This resulted in an increased demand for large-scale export production. Because of labor shortages, however, these demands could not be met directly. Therefore, in line with the expansionism of modern technology noted above, it was proposed to solve this problem through mechanization of the planting and harvesting practices. As a consequence, a further change of the existing farming systems would be required: instead of the traditional practice of intercropping of yam and sugarcane, monocultural growing of yam tubers would become necessary. Special demonstrations were organized on so-called farmers' days in order to propagate these new practices among the farmers. None of these novel conditions, however, had been explicitly defined in any of the project proposals. In this form, however, the technology is obviously inappropriate for the majority of the local farmers. First, they cannot themselves afford the investments and costs of machine planting and harvesting, while no acceptable credit facilities are available either. And second, from their point of view the required change from traditional intercropping to monocropping implies a loss of valuable, indigenous skills, specifically in the field of pest control, and thus entails a potential hazard in terms of an increased susceptibility to diseases.

Given the complexity of the farming system, it is not enough to affirm the desirability of the intended realization for the farmers themselves. Other individuals or groups may experience undesirable conditions as well. A new technology may, for instance, influence the existing roles and goals of men and women. An illustration can be found in a project aimed at improving the production of groundnuts in Zimbabwe (Broerse, 1992). The traditional growing practices for groundnuts are entirely within the province of the women. The project in question, however, requires definite changes of these practices, notably the buying of certain biotic inputs. Consequently, money will be involved and, given the cultural setting in Zimbabwe, growing groundnuts will then become, at least in part, men's business. In this way, a simple "technical" change in the production of groundnuts can have extensive cultural implications, the desirability of which requires explicit attention.

A further significant issue is implied by the temporal aspect of the notion of realization: *how long* can a technological system be expected to be realizable? Appropriate technologies should possess a sufficient degree of robustness: they should not adversely affect the stability of the outputs and the sustainability of the farming systems (Bunders, 1990). The new output should be at least as resilient as the original output to the normal stress and fluctuations to which it is exposed. Pests and diseases, drought or frost periods, and the unavailability of inputs may perturb outputs, but any intended innovation should not increase the negative impacts of such perturbations. At the same time, an innovation should not result in an exhaustion of the basic resources of the farming systems: the biotic and abiotic resources, such as soils, water, and gene pools of plant and animal species; the economic resources, such as capital and labor; and the sociocultural resources such as the organization of labor, cooperation, mutual help practices, and valuable agricultural know-how.

The yam tissue culture project led to innovations that were far from robust. Admittedly, in its first phase the project did result in the eradication of the disease symptoms, while the tissue-cultured clones considerably improved the quantity and quality of the yields. This was definitely a contribution to the stability of the output. Later on, however, in the 1985–1986 season, the region suffered a severe outbreak of the fungal disease anthracnose. The consequences were disastrous. Yam yields have been devastated, and farmers are abandoning the crop. In Barbados, the area under yam has fallen by 75 percent. Anthracnose is now endemic in Barbados and throughout the Caribbean. Apart from this, the advice to focus on monocropping instead of intercropping could be expected to lead to a significant loss of indigenous knowledge with respect to pest control and, thus, to a decreased stability and sustainability of the farming systems in the long run.

It would be wrong, however, to diagnose the anthracnose problem exclusively in terms of "undesirable but unforeseeable hazards and risks," not systematically related to the realization process as a whole. Certainly, it *was* bad luck that the yam variety originally selected for the tissue culture project proved to be especially susceptible to the disease. The severity of the disaster, however, can be positively linked to the particularities of the technological realization. After all, tissue culture techniques focussing on only one variety will lead to a large uniformity in the crops produced. Therefore, if diseases arise, the biological law applies that says that uniformity tends to strongly increase their spread and impact. We have to conclude that, during its intended period of realization, the tissue culture project was certainly not appropriate with regard to hazards and risks.[8]

Ultimately, the feasibility of new agricultural biotechnologies at the farming level should be tested in the field and under local farming conditions. Evaluations show that field tests are often executed under special conditions, such as high rewards in the form of money, inputs, or prices in order to have farmers pay extra attention to their parts in the realization process. Obviously, these tests will lead to unrealistic expectations, when such stimulating conditions are not present or cannot be maintained in the normal farming practices. Hence, success criteria of tests should not be developed without participation of farmers, and the reports of the studies should include a discussion of farmers' doubts and concerns. Discussions with farmers on the suggested changes in agricultural practices will provide valuable insights with respect to their preferences and options. Sometimes field trials are carried out by nongovernmental organizations, who claim to have good relationships with farmers. Recently, these claims have been acknowledged by funding agencies and by several governments. Thus, if we want to learn something from the failures discussed above, one option is to increase the involvement of nongovernmental organizations in the process of realizing new technologies at the farming level. Yet, although a strong relationship with farmers is important, it is not sufficient for executing field trials effectively. Given the division of labor in the realization of biotechnologies, people skilled professionally in setting up these trials are needed as well.

Thus, biotechnological innovations quite generally require drastic changes in the elements of the farming systems. New demands are made on cropping and pest control practices, on credit facilities, on additional labor during harvesting, on traditional gender roles, etc. Can these demands be met at the right moment and are these changes desirable? Part of the appropriateness of the (intended) technology is that such questions are posed, discussed among, and answered positively by, the people and organizations involved in the project.

The Realization of Input Production Systems

Many biotechnological projects in agriculture involve the production of new inputs, such as improved seeds, biofertilizers, biopesticides, products to improve fermentation, etc. In the tissue culture project the production of inputs took place in two stages. In Barbados a tissue culture laboratory was established for producing small amounts of "elite stock." And an "approved growers scheme" was devised for yielding more substantial quantities of virus-free tubers with the help of the elite stock and for distributing these tubers among the local growers. In both stages, a number of crucial material and social conditions needed for a stable and sustainable realization of the products turned out to be unrealizable. In other words, a successful translation of scientific tissue culture results, first to a laboratory in the Caribbean and then to the approved growers, was impossible to obtain.

At the start of the project, not much was known about how the process could be controlled with respect to safety. In the Caribbean, no tissue culture production unit had been established before. Maintenance and multiplication of the virus-free material proved to be difficult. Between 1984 and 1986, delivery of improved material to approved growers completely halted. In 1986 a large proportion of the yam clones was found to be contaminated by bacteria. This setback was due to a breakdown of the electric air conditioning system in the culture room and a generally insufficient standard of hygiene. Subsequently, the tissue culture clones also showed symptoms of virus infections. The cause of the latter infections is still disputed. Some claim that the tissue culture material in the original meristem culture work in the 1970s was virus free, but that poor handling by unsupervised staff (between 1984 and 1986) resulted in reinfection. Others assume that the yam material has always contained significant levels of infection, which the original researchers could not detect because at the time no sufficiently sensitive diagnostic methods were available to them.

It is quite plausible that adherence to good laboratory and manufacturing practice throughout the project would have prevented some of the problems encountered with bacterial infection. If these practices cannot be maintained due to lack of technical infrastructure (electricity) or skills (unsupervised staff), realization of the technology is unfeasible and inappropriate (see also chapter 6). Because of the importance of the continuous availability of electricity in many biotechnological projects, it has been proposed that project designs should routinely include electricity generators, which can be used when the main power fails. Problems with unsupervised staff might be handled by adequate funding support. When this support cannot be realized, technological projects are likely to become inappropriate in the long run. In the yam tissue culture project, for instance, the coincidence between the end of the funding period and the start of the previously described problems was certainly no accident.

The second stage of the project comprised the further multiplication and distribution of virus-free material. Selected approved growers in Barbados could purchase yam "elite" planting material from the tissue culture unit, multiply it, and sell their output to commercial and subsistence growers as "certified disease-free" seed tubers. A precondition for approved growers was that they adhere to a stringent set of conditions and follow precise instructions for growing the materials. These instructions concerned the site of their plots, the planting and cultivating practices, the use of fertilizers and other inputs, and the harvesting and storage techniques. Furthermore, all the yams produced were to be sold for planting purposes only. Thus, disease-free planting material was to be produced by farmers for farmers. In practice, however, it proved to be difficult to meet these stringent conditions. It became evident that, in seeking higher profits, approved growers multiplied the improved varieties conventionally in their fields rather than returning to the production unit for new disease-free stock. Thus, the realization of the scale-up of the production from the tissue culture unit to the approved growers appeared to be problematic. Conversely, the fact that many of the approved growers did not purchase their planting material from the tissue culture unit aggravated the financial problems that already plagued it by that time.

Lack of quality control in the input production systems is a serious problem in many countries. Cases have been reported of inoculants that were sold for yield improvement but in fact contained nothing but soil and dust. Also propagation and distribution of technologically improved products to resource-poor farmers in neglected areas are often extremely difficult, hampering the realization of an intended innovation. In these respects, the tissue culture case is instructive. It teaches governmental, nongovernmental, scientific, and industrial organizations what kinds of requirements have to be met for appropriately realizing agricultural technologies, especially in developing countries.

Dependence on Science

The use of scientific knowledge, skills, techniques, and materials is one of the key features of modern technology. This also applies in the case of agricultural biotechnology. Up to now, research and development in biotechnology has primarily been executed in developed countries. If developing countries wish to use biotechnological results, they have to depend on the developed countries, which probably have research priorities that differ from theirs. Thus, an important, normatively relevant question is: should developing countries attempt to realize their own capacity in biotechological research, or is it more desirable that research remains concentrated in the developed countries?

This issue also played a role in the yam project. Apart from the direct aim of finding out whether tissue culture on yam was realizable at a laboratory scale, more general objectives of the project were the technology transfer of tissue culture methods and the buildup of an independent research capacity in the Caribbean. The project involved the active participation of local researchers and resulted in the transfer of key biotechnological skills and resources. Thus, with respect to the issue of a reduction of international dependence, the project was quite appropriate.

More generally, it is sometimes suggested that developing countries should now try to catch up, since some opportunities are still present that are expected to disappear in the near future (Perez and Soete, 1988). Currently, the buildup of local research capacities would be relatively easy because, among other things, most biotechnological knowledge is still publicly available. Thus, this argument hinges on the account of technology—which was mentioned in section 7.2—as a matter of straightforward application of available knowledge. On the basis of our analysis of technology and of our case study of the yam tissue culture project, we think that the view that biotechnology can be realized relatively easy in other contexts (for instance, the context of resource-poor farmers) *because* it is publicly available, is highly unrealistic. First, most of the research is neither developed nor structured from the viewpoint of resource-poor farmers in developing countries. Second, the results are published almost exclusively in highly specialized journals, which are quite inaccessible to nonspecialists because of their extensive use of technical jargon. Third, studies of experimental replication show that in general it is difficult (if not impossible) to reproduce and use experimental results on the basis of published texts alone (cf. Collins, 1985).

In order to be able to successfully realize biotechnological research in developing countries many heterogeneous conditions need to be fulfilled. This conclusion is shared by many policymakers. It has led to extensive debates and calls to design policies that deal with this problem (see, e.g., Broerse and Wessels, 1989; Bukman, 1989). In the Netherlands, the journal *Biotechnology and Development Monitor* has been established as a result of these debates. The aim of the *Monitor* is to bridge the gap between the research results of the specialists and the biotechnological needs of developing countries. It presents all kinds of relevant information on biotechnology and developing countries in a popularized form to interested institutions and individuals. It also publishes overviews of present biotechnological research programs from a broad variety of institutions. In order to set up a biotechnological research capacity, however, much more than access to information is needed. As is implied by the analysis in this chapter, appropriate research capacity building requires, among other things, the making and executing of comprehensive policies in both developing and developed countries.

The Commercialization of Biotechnological Research

A major implication of the discussion so far is that appropriate technology for small-scale farmers in developing countries should be based on a thorough knowledge of their farming systems. In practice, acquiring such knowledge is quite difficult. Fieldwork might be done, but it will often be hampered by lack of funds, cultural differences between farmers and researchers, and insufficient communication skills. These problems are issues for debates and political agendas. Adequate policies and committed people who are interested in improving the living conditions of the rural poor are essential to finding appropriate solutions.

Moreover, acquiring the knowledge required for realizing new or improved technologies is by no means necessarily desirable. Again, one should keep in mind that doing scientific research is not just a matter of having ideas but also involves the availability of the relevant growing materials from developing countries. On this issue, an important and critical debate has arisen about the "farmers' rights" with respect to these materials. The starting point is that it is farmers who have contributed most to the creation, conservation, and exchange of many valuable species. Their germplasm conservation activities are based on knowledge and expertise passed on from generation to generation. Therefore, many people feel that gathering plant materials improved through farmers' expertise is undesirable if used by powerful commercial research organizations in universities or industries, which are not interested in improving the production systems of these farmers. For instance, Mushita, who works for ENDA (Environment and Development Activities)-Zimbabwe, claims that farmers' plant genetic resources and skills are now exploited during expeditions for collecting promising new material, without recognition or compensation for these farmers (Mushita, 1992). In the same vein, Hobbelink concludes (in Brouwer, Stokhof, and Bunders, 1992, 91–93) that it is unjust that the freely available plant genetic resources from developing countries are taken to industrialized countries, where they can be processed and patented, and thus will not be freely available any more. At present no formal or legal framework exists for recognizing and rewarding farmers' efforts with respect to the conservation of plant genetic resources. On several occasions, a plea has been made to acknowledge these rights of farmers. Some suggest the formation of a mandatory fund. Others propose compensatory measures in the form of a stronger focus on the needs and wishes of resource-poor farmers. Thus, at stake here is a significant normative issue, which should be dealt with in assessing the appropriateness of the relevant technologies.

Organizing Appropriate Systemic Interactions

Given the systemic character of technology, organizing the interaction, coordination, or integration of the different components is a crucial requisite for the feasibility and desirability of its realization. The tissue culture project was organized step by step. First the biotechnological research was carried out. Subsequently, a design was made for the production and distribution system. Finally, the issue of the use of the product by ordinary farmers was tackled. The disadvantage of such an approach is that during the research stage, an adequate anticipation of problems in the context of production and use is impossible. As we have seen, many problems that arose in the course of the project were due to such a lack of interaction and coordination.

Therefore, a much preferable alternative is to incorporate the issue of the organization of systemic interactions as a regular topic at the start of technological (research) projects. Even so there is clearly no guarantee of success. The problems to be faced are large and heterogeneous. They derive from, among other things, the uncertainty of future developments, the dynamical character of the different contexts, and the cognitive, professional, and cultural differences between actors who have no tradition of communicating with each other. Each group possesses its own specific expertise but lacks other necessary knowledge and skills. Obviously such problems will impede the appropriate realization of the intended technology. This situation is, however, not typical of biotechnological projects for small-scale farmers but is characteristic of all technological innovations. Much can be learned therefore from investigations of innovative processes. Analyses have been made of the ways in which entrepreneurs have succeeded in implementing innovations within their organizations. Other analyses show how industry influences public research, and how other—less powerful—groups have successfully written their priorities on the public research agenda (Bunders and Leydesdorff, 1987; Bunders, Stolp, and Broerse, 1991, 81–96).

Results of such studies have been incorporated into a model for appropriately organizing systemic interactions in farmer-led biotechnological innovation processes: the *interactive bottom-up approach*.[9] This approach avoids technology push by drawing on the needs, wishes and knowledge not only of scientists, policymakers, and expert consultants but also of end users and organizations that represent them or work with them. One of its central features is the use of two different but closely cooperating teams: a formally operating, interdisciplinary team that aims to scrutinize the gap between the claimed potential and the actual realization of the intended technology; and an informal team, consisting of people sharing the same commitments, to investigate and establish whether or not the realization of the technology is considered feasible and desirable by all the people involved. This model has been used in a number of different countries. The design of policies that take into account the entire technological system proves to be a vital requisite. Developing countries differ widely in their willingness to design such policies. For example, the governmental focus on resource-poor farmers in Zimbabwe is much stronger than in Pakistan (Broerse, 1990; Bunders, 1990). The realization of appropriate agricultural biotechnology is therefore less likely in Pakistan than it is in Zimbabwe. Nevertheless, our experience indicates that the organization of systemic interactions aiming at appropriate biotechnology for small-scale farmers in developing countries can be fruitfully tackled with the help of the interactive bottom-up approach.

7.6 CONCLUSION

In this chapter, a new framework has been proposed for dealing with ethical or, more generally, normative issues surrounding technology, which has been specified and illustrated by reference to the case of small-scale farmers in developing countries. Basically, the framework consists of the key features of technology explained in section 7.3, together with the notion of appropriate technology, as defined with the help of the central questions posed and discussed in section 7.4. The specification and illustration of the framework uncovered a whole range of normative or ethical issues connected with the (intended) realization of agricultural biotechnologies in developing countries. The aim has been not just to contribute to a theoretical understanding of the realization of technologies but also to offer an approach that can be helpful in attempts to proceed towards greater appropriateness. Thus, we see our account as a part of, and ourselves as participants in, the technological systems in question. In this respect our approach appears to be consistent with Walker's account of the role of ethicists as "architects of moral space" and as "mediators of the conversations taking place within that space" (Walker, 1993, 39–40).

We have also stressed that in practice, appropriateness will be a matter of degree. Certain technological systems may be more appropriate than others, and certain aspects of a technological system may be more appropriate than other aspects. Consequently, the central normative question is how the degree of appropriateness might be increased. As we have seen, due to the heterogeneity of technology itself, the possible resolutions of a normative issue are heterogeneous as well. One may look for technical solutions, forbear from realizing the technology, design and implement new price policies, aim to transform gender roles, enforce professional laboratory standards, or fight the legacy of colonialism. More generally, the identification and discussion of less appropriate (aspects of) technologies by the people involved may have various outcomes. It may induce suggestions for changing particular components of the technological system. It may lead to the conclusion that the technology in question is highly inappropriate and that other technical or nontechnical approaches should be considered. Or, it may result in compliance with the normative deficiencies, as more appropriate solutions cannot be found or obtained.

Of course, applying the proposed framework is not sufficient for obtaining more appropriate technologies. The inertia of existing material, psychological, social, and cultural realities may simply be too large. Nevertheless, the practical relevance of the framework depends on committed attempts toward greater appropriateness. Looking at the experiences gained so far, we see that relevant changes are possible in terms of different project setups, different policies, and different institutions. Whether or not these particular interventions will really result in more appropriate technologies needs to be established in the near future.

CHAPTER 8

Philosophy:
In and about the World

8.1 INTRODUCTION

In this final chapter, I want to reflect on the kind of approach I have taken in the preceding studies. According to the subtitle of this book, these studies are "philosophical." What I mean is that they have resulted from employing skills and taking approaches that I would like to label "philosophical." In the subsequent sections, the nature of this philosophy will be explained more explicitly and in more detail. What I propose is a threefold characterization of philosophy. First, philosophy is *theoretical*, in the two senses of being both explanatory and interpretative. I will discuss this aspect in section 8.2. The second aspect, to be dealt with in section 8.3, is the *normative* character of philosophical approaches. Finally, there is the *reflexive* aspect of philosophy, which I will examine in section 8.4. To be clear, when I characterize philosophy as theoretical, normative, and reflexive, I do not mean to offer a rigorous definition in terms of necessary or sufficient conditions. The characterization rather bears upon a nonlocal pattern that can be recognized in all kinds of philosophical conceptions, approaches, and debates. As a pattern, however, it should exhibit all three aspects of philosophy. For instance, a separate claim such as "application letters ought to be handwritten" does not count as a philosophical claim just because of its being normative.

In conceiving of philosophy in this manner, the intention is not to surround it with high disciplinary walls in order to protect it from potentially hostile outside invaders. I do think, however, that an approach that is theoretical, normative, and reflexive embodies specific skills and is able to provide valuable insights. This does not imply that the prevailing conceptions of philosophical theory, philosophical normativity, and philosophical reflexivity can be uncritically adopted. By situating itself as being at once in and about the world, the present account aims to go beyond the more familiar conceptions of philosophy based on such oppositions as those between discovery and justification, the a posteriori and the a priori, the particular and the universal, or the social and the rational.[1]

The subsequent metaphilosophical discussions are "meta" in the literal sense of following up on the philosophical chapters. This is as it should be. In this case, the systematic development of metaphilosophical views will be guided by a sensitivity to the complexity and difficulty of formulating and answering particular philosophical questions in a detailed manner. Retaining this order will counteract the tendency—visible in quite some recent approaches—that metaphilosophies either get stuck in merely programmatic, "grand" views or lead to an unfruitful, self-sufficient discourse at a quasi-autonomous metalevel.

8.2 PHILOSOPHY AS THEORETICAL

Theoretical philosophy of science and technology, as I see it, endeavors to expose and examine structural features that explain or make sense of nonlocal patterns in the practices, processes, and products of science and technology. In this conception, "theoretical" is not primarily contrasted with "practical" but rather with "empirical." The terms "process" and "product" have been added to emphasize that theoretical philosophy of science and technology includes the study of method and knowledge claims. It will also be clear that exhibiting and analyzing nonlocal patterns is not merely a matter of "empirical generalization" but requires in its turn a theoretical perspective and theoretical work.

Let me first give two examples from the preceding chapters to illustrate such philosophical explanation and interpretation. In section 2.2 I have put forward the notion of the description of the material realization of experiments on the basis of processes of communication and division of labor between theoretically informed experimenters and theoretically non-informed laypersons. As I pointed out, this specific operationalization of the process of material realization is not meant to be a straightforward description of empirical patterns in experimental practice but rather to characterize some structural features of experimental action and production, including their relationship to theoretical description. In this way it explains both the interaction between and the relative autonomy of experimental manipulation and theoretical interpretation. Furthermore, although the account is not meant as descriptive of all scientific practice, this does not imply that it cannot be empirically supported. In fact, this particular theoretical account of materially realizing experiments includes the conditions under which it may be practically substantiated, namely in cases of division of manipulative and theoretical labor.[2] In chapter 2 I have provided some illustrations of such cases, taken from historical and sociological studies of experimental practice. Apart from this, employing explanatory theoretical concepts that are not primarily descriptive and yet empirically supportable is quite common in other sciences. Thus, generally speaking, explanatory theoretical philosophy is just as legitimate, and can be equally informative, as any other theoretical science.

As a second example, consider the realist ontology for experimental and observational science proposed and discussed in chapter 4. There I argued that the meaning of "a reproducible experiment" transcends the meaning of any pattern of actually performed reproductions. In this sense, reproducibility is not simply a descriptive term. Nevertheless, the claim that an experiment is reproducible can be empirically supported by the occurrence of a nonlocal pattern of successful reproductions. I further argued that, if a term from a theoretical description of a reproducible experiment refers, it is about a persistent potentiality of a human-independent reality. Thus, the ontological interpretation in terms of persistent potentialities purports to make sense of the successfully made (implicit or explicit) claims on reproducibility, by which reproducible experiments transcend the contexts of their particular, local realizations.

More broadly, any account of science or technology will make use of certain comprehensive philosophical assumptions. These assumptions constitute, in a hermeneutical sense, the meaning that is attached to science or technology within the account in question. The support that such interpretative philosophical claims may derive from the existence of nonlocal patterns of scientific or technological practice will generally be more indirect than in the case of theoretical explanations. If successful, philosophical interpretations make sense of certain nonlocal patterns in a coherent and illuminating manner, and in this way they may be said to be adequate to the relevant practices. Apart from this, the plausibility of philosophical interpretations will depend on other features, such as their logical coherence, conceptual clarity, heuristic fertility, sociocultural significance, and normative stake.

The Inadequacy of Metaempiricism

Doing philosophy as a theoretical—that is, primarily an explanatory or interpretative activity—contrasts to empiricist or empiricist-minded approaches. The lattter can be found quite frequently in constructivist studies of science and technology, as we have seen in chapter 5. The ethnomethodological (or ethnographical) program is the most radical. According to Lynch, Livingston, and Garfinkel, "the overriding preoccupation in ethnomethodological studies is with the detailed and *observable* practices which make up the incarnate production of ordinary social facts."[3] Comparable empiricist views are quite widespread in recent science and technology studies, ethnomethodologists being only among its most persistent adherents. Thus, Callon, Law, and Rip (1986, 14–15) claim that they merely follow technoscientists "with neither passion nor prejudice," and that they are simply "coolly describing" what they see. And even in 1992, Callon and Latour state that "in Paris and Bath we all agree that the touchstone of any position is its empirical fruitfulness," and that in debates among students of science and technology, the empirical (dis)agreement is "the only one that really matters" (Callon and Latour, 1992, 345 and 352).

From such empiricist metaperspectives, it is either claimed or implied that (theoretical) philosophy has actually lost its right of existence. The best philosophers can do, so it seems, is to close their offices and turn to the empirical study of science and technology. In that manner they can then contribute to the development of authentic accounts of what is really going on in the day-to-day work of the practicing scientists or technologists. To be frank, I should say that—from an autobiographic point of view—I find all this empiricist metatalk rather embarrassing. I started my philosophical studies at the time of the decline of the logical empiricist tradition. I then welcomed the subsequent postpositivist and constructivist approaches. Should I, after these significant developments during the last decades, now find myself back in the empiricist bandwagon?

Fortunately, this is not at all necessary, since the constructivists' metaempiricism meets with a number of problems that are as fundamental and as unsolvable as those that confronted their predecessors, the logical empiricists. For one thing, there are the many convincing criticisms that have been leveled against empiricist interpretations of science in general. Since they are well known, I will not bother with repeating them.[4] One important point needs to be added, though. This is the rather totalitarian character of the constructivists' metaempiricism. After all, the empirical study of local episodes is claimed to be the *only* way to learn something about science or technology. Compared with this claim, a philosophical empiricist such as Van Fraassen is relatively modest.[5] He merely claims that the representational status of scientific theories should only be evaluated in terms of their empirical adequacy. Next to representation, however, Van Fraassen allows (and describes) other goals of science—such as explanation—that cannot be analyzed and evaluated exclusively in terms of empiricist notions. A consequence of the radical empiricism of the constructivists is the self-fulfilling character of some of their claims. This applies in particular to the locality thesis (see chapters 4 and 5). To put it in a straightforward manner, if one studies merely disconnected, local practices and does not look for nonlocal patterns and structural, explanatory features, one will, of course, find that science is no more than a loose collection of disconnected, local practices.

Furthermore, it is not difficult to demonstrate that the constructivists do not live up to their proclaimed methodological standards.

Indeed, many claims by recent constructivist students of science and technology go far beyond what can be "coolly observed" in empirical studies of scientific or technological practice. Claims that "facts are no more than social constructions" or that "there is no essential distinction between human and nonhuman actors" can be illuminated by, but by no means derived from, empirical studies. These are full-blown philosophical claims, which for that reason require a philosophical discussion.[6] As a matter of fact, the more sophisticated advocates of such claims have unavoidably found themselves engaged in interesting and fundamental philosophical debates. As a result, within constructivism an unsolved tension can be seen to obtain between empiricist creed and heterogeneous practice.

Thus, the constructivists' empiricism functions not so much as a consistently applied research method but rather as an inadequate self-image. The following is just one instance of this inadequacy. Within constructivism the notion of "reconstruction" is frequently used to differentiate philosophy from empirical approaches. The (incorrect) suggestion is that empirical approaches straightforwardly capture the "real-time" development of science or technology, whereas theoretical explanations or interpretations are no more than "post hoc reconstructions." The simple truth is, however, that *all* accounts of science and technology are re-constructions, which start from specific premises and perspectives. Different accounts do, of course, differ, but their difference lies in *what* they reconstruct and *how* they do it.

In fairness, not all constructivists endorse the extreme empiricist claims discussed so far. As one case, Bijker's Social Construction Of Technology (SCOT) model exemplifies a more relaxed approach. It explicitly agrees that there is more to technology than what separate, empirical case studies can reveal. What is needed and possible, so it is argued, is to search for descriptive empirical generalizations of the development of technology, on the basis of the SCOT model that is claimed to be universally applicable. Bijker (1990, 178) concludes that "the resulting theory can thus be said to be 'grounded' in empirical research." In my view, attempts to construct nonlocal patterns (as I prefer to call them) are perfectly sensible. However, as will be clear from the above discussion, they do not exhaust the theoretical study of technology. Consider for instance the SCOT assumption that the social construction of technology is accomplished through clearly identifiable social groups that attribute an equally clearly identifiable meaning to the relevant technological artifacts. Obviously, this assumption implies a specific theoretical view of what society is and how it operates. This view precedes the empirical case studies, and it may, therefore, be rightfully contested on grounds that are independent from the particulars of the relevant cases.[7] For reasons like these, the residual empiricist metatalk about "empirical generalization" and "empirically grounded theory" may easily mislead, and thus it had better be dropped.[8]

What Is the Explanandum or Interpretandum of Theoretical Philosophy?

Having said all this, it is time to turn to the other side of the issue and give the empirical (historical or sociological) approaches their due. When I propose philosophy as explanatory or interpretative, an immediate question is, of course, what the *explanandum* or *interpretandum* is. My answer to this question is rather straightforward: anything that plays a role with respect to the practices, processes, and products of science and technology may be taken into account in philosophical explanations and interpretations. But of course, this answer is "straightforward" only in a general sense. In concrete cases a careful study of the particular *explananda* or *interpretanda* is required. Philosophers must possess a good deal of insight into the practices, processes, and products of science, especially in view of assessing the relative significance of a philosophical explanation or interpretation. For instance, does the explanation or interpretation in question apply to all stages of scientific research or only to a particular one? In order to be able to answer such a question, an interactive and cooperative attitude towards (descriptive and "generalizing") historical and sociological studies is required. As a result, just as is the case in other sciences, such an interaction and cooperation between theoretical and empirical approaches can be expected to lead to a more adequate and more comprehensive understanding of science and technology.[9]

Despite its seeming triviality, my answer to the question of the *explananda* or *interpretanda* of theoretical philosophy has some important consequences. It is, for example, critical of any a priori fixing of "basic units" for the philosophy of science. Since any explanation starts from a certain "question context," different questions may require different "basic units." For instance, choosing individual scientists as basic units (see Giere, 1988, ch. 6), may well be a good starting point for discussing the structure of decision making in science, but it may be utterly inadequate for developing an interpretation of the relationship between knowledge and reality.

A more general and far-reaching implication of the above point of view is that it induces us to move beyond the received Kuhnian dichotomy of the "real-time" scientific discoveries on the one hand and the merely didactic use of textbook presentations on the other. After all, both play a significant role in scientific practice, and there are no good reasons why one of them could be a priori disregarded as not or less relevant for acquiring an adequate view of science. The significance of this point has been explicitly put forward by Nickles in an illuminating manner:

> Science transforms itself by more or less continuously reworking its previous results and techniques. To miss the dynamical, self-reconstructive nature of scientific work is to miss the extent to which scientific inquiry is a bootstrap affair. I shall call non-reconstructive views of science *single-pass* or *one-pass* models of scientific inquiry. . . . In ignoring the self-transforming character of scientific work, one-pass models can hardly avoid committing the *genetic fallacy*, the mistake of thinking that the conditions of origin determine forever the character or "essence" of a thing.[10]

The importance of reconstruction and the necessity of a multipass account can also be found in the preceding studies. In chapter 3, for instance, I have emphasized the transformational character of all processes of scientific knowledge production. As we have seen, the construction of a new theory by means of the correspondence principle can be described as a transformation of an older theory, which then acquires a reconstructed meaning by being reinterpreted from the perspective of the new theory. In the same vein, the discussion of experimental reproducibility in chapter 2 pointed out the significance of various kinds of reconstructive, stabilizing procedures, such as experimental replication or standardization. More in general, the philosophical explanations and interpretations in these and the other chapters clearly assumed a multipass account of the development of science and technology.[11]

8.3 PHILOSOPHY AS NORMATIVE

Traditionally, philosophy of science and more generally epistemology have been conceived as (at least partly) normative disciplines. The task of philosophers would be to formulate criteria that make it possible to analyze, assess, and possibly criticize scientific methods and scientific knowledge from a point of view that is either outside science or at least more comprehensive than science. What has passed the test of philosophical criteria can genuinely be labeled "good" or "justified" science. In the course of time, numerous and rather dissimilar criteria have been proposed. The differences concern not just the specific content of the criteria but also what they are supposed to achieve. Verifiability, falsifiability, verisimilitude, coherence, explanatory and predictive power, problem solving capacity, and unconstrained communication denote some of the most influential proposals. Meaningfulness, demarcation, correspondence to reality, rationality, and progress constitute some well-known goals that are supposed to be reached by applying the relevant criteria.

Three observations are in order with respect to this brief sketch of the framework of traditional, normative philosophy. First, even when the philosophical criteria are located "outside" science, this does not imply that the normative points of view are entirely independent of the development of science. After all, the normative philosophers would certainly be embarrassed if their criteria appeared to be generally violated. Therefore, most of them "have in mind" some exemplars of what they take to be good science and good scientists. These exemplars, which are supposed to illustrate the advocated philosophical criteria, get a paradigmatic function for all of science. Then, from a quick look at one or two other episodes, it is concluded that, although of course some irrational deviations occur, science "mostly" conforms to the philosophical criteria in question.

A second observation concerns the scope of the normative philosophical views. At least in a number of important cases, the significance of the normative proposals is claimed to be much wider than their applicability to science proper. The assessment and critique of science is often continuous with an assessment and critique of society. The norms put forward for establishing the rationality of science are taken to be equally relevant to rationally organizing society. This is of course evidently true in the case of approaches that primarily intend to embed a theory of (scientific) knowledge within a social critical theory of society (see, for instance, Habermas, 1978). But it also applies to major analytical philosophies. For instance, reading the 1929 manifesto of the Vienna Circle, "The Scientific Conception of the World," immediately reveals the strong social critical roots and aims of the logical positivist approach to science (Neurath and Cohen, 1973, 299–318). Also in the case of Popper, significant links exist between his philosophical and his social and political views. For example, in his memories of Neurath, he claims that Neurath and he disagreed on almost all matters, except on "the view that the theory of knowledge was important for an understanding of history and of political problems" (Neurath and Cohen, 1973, 56; cf. also Krige, 1978).

A third feature of the normative philosophical criteria is their global nature. Both the epistemological and the social critical approaches lead to assessments and normative recommendations of science "in general." Moreover, many philosophers situate their assessments and recommendations within the so-called context of justification, thus implying that they are not directly relevant to the context of discovery. As a consequence, the question of the realizability and effectiveness of the proposed norms in concrete scientific practice is not systematically studied, let alone satisfactorily answered, by traditional philosophers. In Fuller's words, they circumscribe the sphere of the normative so that the philosopher presents herself as *evaluating*, but not actually *directing*, the development of science. This gives the philosophy of science its apolitical sense of critique, which is reminiscent of evaluation in aesthetics. (Fuller, 1993, 37)

Naturalist Approaches and Their Problems

The project of the normative evaluation of (scientific) knowledge has been fundamentally criticized by advocates of a naturalist epistemology and philosophy of science. In recent times this approach has attracted an increasing number of adherents. The concept of naturalism itself has been defined in various ways, but in the context of the present discussion, the following general characterization of (epistemological) naturalism provides a good starting point:

> Proponents of naturalized epistemology maintain that human knowledge is a natural phenomenon to be studied by the same scientific techniques that we use to study any other aspect of nature.[12]

By now different naturalist schools have already emerged. Some single out a specific discipline (evolutionary biology, neuroscience, psychology, history, or sociology) as being especially appropriate to the study of (scientific) knowledge, while others hold the more liberal view that any scientific approach that may supply relevant information about the production of (scientific) knowledge is welcome. The rise of naturalism has been stimulated in part by the appeal of the results of particular approaches that are seen as especially impressive and fruitful. Well-known instances are the historical and the artificial intelligence approach to (scientific) knowledge. In part, naturalism has also been supported because it is claimed to be a viable alternative that does not suffer from the intrinsic and unsolvable problems clinging to the traditional normative approaches.

The general naturalist criticisms of normative epistemology and philosophy of science are straightforward. They focus on two major issues. The first is that, even after centuries of great effort, philosophers have not been able to put forward any generally acceptable normative criteria. All criteria proposed so far have been continually disputed, on the basis of claimed fundamental inadequacies. In this sense Quine, to mention one influential naturalist in epistemology, concludes from the continuing failure to deduce theoretical knowledge claims from, or translate them into, an observational basis, that the project of normative justificationism as such should be given up (see Quine, 1985). Instead, Quine advocates the behaviorist psychological approach as the royal road for investigating the main epistemological problem, which—according to him—is how observational inputs are being processed into reliable theoretical knowledge.

A second major point of criticism, pertaining more particularly to the philosophy of science, has been that the various norms proposed by philosophers are frequently violated by the scientists in question. In other words, apart from their philosophical inadequacies, these proposals also suffer from serious empirical inadequacies. The paradigmatic examples of good science are claimed to be either not paradigmatic at all or not even conforming to the theories of the normative philosophers themselves. Thus Kuhn, the most well-known historicist naturalist, has sharply criticized Popper's normative, falsificationist account of scientific method and scientific knowledge (see Kuhn, 1970a). As an alternative, he argues for a major role for the historical study of scientific practice, with the purpose of achieving an adequate descriptive understanding of scientific development.

What to think of this debate between normative and naturalist approaches from the perspective of the philosophical studies of science and technology presented in this book? On the one hand, I do agree with the naturalists that the traditional conception of epistemology and philosophy of science is fundamentally flawed. In fact, this agreement provided the motivation and starting point for framing the preceding chapters in the way I did. On the other hand, the correctness of its criticism does not imply that the naturalist alternative is acceptable as such. In order to demonstrate this, I will make a number of critical observations concerning the naturalist approach. In doing so, I focus on naturalist philosophy of science, although some of the points are also relevant to naturalist epistemology.

In a certain sense, the move from a normative to a naturalist approach to the philosophy of science is rather odd. Remember that one of the basic naturalist criticisms is that all attempts at specifying criteria with which to assess what science or scientific method really is have failed dramatically; and remember also the other main criticism, saying that all proposed epistemological or methodological norms have turned out to be inapplicable, because actual science appears to be a complex and variable activity that simply cannot be fitted into the philosopher's straitjacket. So, what exactly is the naturalist advice "to study science scientifically" supposed to mean? The only cogent answer is: we don't know![13]

To get an idea of the far-reaching consequences of this point, consider the following argumentation. Since we do not have the disposal of a demarcation criterion to tell us which approaches may be legitimately employed in the naturalist study of science, a reasonable way to proceed might be to have current practice tell us what counts as a scientific approach. Unfortunately, this move will not solve the problem either. For instance, in the Netherlands and in Germany, all research institutionalized and practiced in governmentally funded universities clearly counts as *wetenschap* or *Wissenschaft*, and thus should qualify as scientific. Consequently, in this context, academic philosophy, whether it is naturalist or normative, is and always has been "scientific." In this sense even the most traditional a priori philosopher may easily endorse the claim that real philosophy should be practiced scientifically.[14]

It will be clear, though, that this is not what most naturalist philosophers have in mind. Therefore, what actually happens is this. In their—quite understandable—search for a more positive content of their program, they sacrifice its consistency by making cryptonormativist prejudgments. Consider for instance the above quotation of Brown. Even though Brown is relatively liberal in allowing for a certain latitude in the choice of the sciences that are thought to be suited for the study of science, he has also prejudged the issue at stake in an essential way. This is shown quite clearly in his statement that human knowledge is a "natural" phenomenon, implying that for a naturalist science of science, natural science remains *the* model of "good" or "legitimate" inquiry (which is, of course, already implicit in the name "naturalism"). But it is more than obvious that this claim presupposes a global normative appraisal of a kind that is close to the assessments made by many normative philosophers of science.

This appraisal, however, is by no means a "natural phenomenon" that is generally agreed upon. In fact, it is and has been continually contested by a substantial group of scientists and philosophers who have argued for the essential distinctness of natural and human reality, of nature and culture. Logically, the situation is simple enough. Naturalists are right to stress that there is no ahistorical, Archimedean point outside science from which we might a priori justify scientific method and scientific knowledge. However, to conclude from this that therefore philosophy and science "are in the same boat," as Quine (1985, 39) does, is to commit a logical mistake. After all, "legitimizing" science by means of normative philosophical evaluations from (admittedly contingent) "outside" or "more comprehensive" perspectives remains perfectly possible. Indeed, the latter precisely captures what scientistic naturalists are doing all of the time (cf. Hookway, 1988, 61–63).

Because this point has gone relatively unnoticed so far (something which I find hard to understand, since it appears to be so evident), let me give one more illustration. In his article on naturalized philosophy of science, Giere poses the question whether his approach implies "that 'creation theory' is as good as evolutionary theory?" His reply to this is: "No more than that it implies that prayer is as effective as penicillin for curing infections" (Giere, 1985, 341). This answer is questionable, for two reasons. First, his example could hardly have been more unfortunate. After all, it is well known that in the area of medicine what is effective and what is not is an extremely delicate matter. In a lot of cases, scientifically respectable treatments (including administering penicillin) turn out not to work; frequently medical cures work but no one knows why; and regularly scientifically obscure or disregarded therapies (including prayer) work perfectly well. Thus, empirically Giere's claim can easily be refuted. More important, however, is the implied normativity. Apparently, science and pseudo- or nonscience can be demarcated by the normative criterion of effectivity! "Good" or "legitimate" scientific knowledge should be "effective." On the basis of this analysis, I cannot but conclude that, with respect to normativity, the only difference between a normative and a naturalist approach appears to be that the former undertakes an explicit but unsuccessful attempt at providing a convincing a priori argument for its normative proposals, whereas the latter has been relatively successful in keeping its (similar) normative presuppositions implicit and hidden.

Against a Scientistically Truncated Normative Naturalism

More recently, some naturalists have proposed a "normative naturalism," which includes not only a descriptive or explanatory but also a normative component.[15] Thus, Giere (1988, xvii) now claims that his view "is thoroughly naturalistic, requiring no special type of rationality beyond the effective use of available means to achieve desired goals." Central to the position of normative naturalism is the view that categorical epistemological or (meta)methodological imperatives do not exist. All normative criteria should be construed as hypothetical imperatives: *if* one wants to reach such and such a goal, then one should act according to such and such criteria. Depending on their context of use, these criteria may be more or less effective and reliable. The basic claim of normative naturalism, then, is that establishing the degree of effectiveness and reliability of a particular criterion is a normatively neutral, empirical matter. In other words, what ends are chosen in science and technology is a normative and contingent matter, whereas whether or not a certain method is effective and reliable in achieving a given end is purely a matter of empirical evidence (see Laudan, 1987, esp. 24–25; Brown, 1988, esp. 64–74).

Now, the question to be discussed is how the preceding philosophical accounts of science and technology relate to such a normative naturalism. Roughly, my view of philosophy appears to be closer to normative naturalism than it is to normative justificationism. I agree that the latter program has systematically failed to make good its promises. Moreover, as explained in the preceding section, I take seriously the naturalist requirement that factual information about science and technology is necessary for constructing philosophical explanations and interpretations and for evaluating their plausibility. Also on this count, most of the normative proposals by traditional philosophers have failed. Finally, the claim that scientific norms have no unconditional applicability is in line with the arguments put forward in the preceding chapters. This is especially apparent from the discussions of the intrinsic connection between knowledge and power. For example, the applicability of the norm that experiments should be reproducible is conditional on the desirability of realizing the material and social conditions that are necessary for creating and maintaining stable experimental or technological systems. A clear illustration of this can be found in section 6.3, in the case of the failure of the (experimentally successful) eradication of boll weevils in the open field, due to a lack of cooperation on the side of the involved farmers. In chapter 7 the same point has been developed more systematically into the notion of the appropriate realization of technological projects.

At the same time, however, the line of argumentation in these chapters entails a significant disagreement with the approach of the normative naturalists. The point is clearly demonstrated in the boll weevil case and in the yam tissue culture project. Establishing scientific or technological effectiveness and reliability is not simply a matter of adducing "neutral empirical evidence," because the realization of this evidence requires a normative commitment from the people involved. As a consequence of their employment of the notion of instrumental rationality, the normative naturalists wrongly stick to a view in which knowledge and power can be separated. According to this view, power, be it legitimate or repressive, may play a role in determining or changing the ends. But given these, it is normatively neutral scientific or technological knowledge that should decide the issue of what are the most effective means for achieving the ends. In a variation on a famous phrase of Habermas, this point of view may be called a "scientistically truncated normative naturalism." Its main defect is that it has not consistently worked through the consequences of the fact that producing scientific and technological evidence is not a matter of having "ideas" floating high above the world but rather a matter of *realizing* these ideas in the world. This implies that the effectiveness and reliability of scientific or technological means is intrinsically connected to questions of material and social feasibility and desirability.

In order to illustrate this point, I will briefly discuss an issue in the area of "artificial intelligence." As I noted above, the claimed successes in this area have been instrumental in promoting naturalist views in the first place. More in particular, consider the question of whether computers are effective and reliable chess players. This question is interesting for two reasons. Designing and testing chess programs has been and continues to be an important topic in artificial intelligence research. Moreover, regularly general and extensive claims concerning the future potential of artificial intelligence have been vindicated by referring to the "remarkable successes in the area of chess playing."[16]

The usual operationalization of the question of the effectiveness of chess computers goes by translating it into the question of whether computers are better chess players than humans. Then it is argued that the latter question can be answered positively by simply looking at the empirical evidence. After all, isn't it the case that at present the best chess computers win of the large majority of human players? Thus, artificial intelligence programs embody a neutral and effective scientific means to the socially defined end of successfully playing the game of chess.

My criticism of this particular instance of instrumental rationality is in line with the comments I made above. Also in this case, instrumental effectiveness is essentially dependent on a concomitant social effectiveness. Before we can answer the question whether computers are better than humans in playing chess, we must answer the question whether computers do play chess at all. And the crucial point is that a (positive or negative) answer to the former question presupposes a *transformation* of the meaning and practice of playing chess. It is only if we define chess as, roughly, the activity of "displaying" well-defined token moves, alternately by white and black, resulting in the well-defined outcomes of win, lose, or draw, that computers are capable of playing chess at all.[17] This definition, however, implies an obvious reduction of what is involved in human chess playing. What has been left out is not only the bodily performance of the moves (which requires considerable concentration and ability when one is running out of time) but also the psychological and social skills that are frequently and legitimately employed. These skills make their own contribution to the complexity and the fun of playing the game. For example, many games have been won by noticing a lack of self-reliance on the part of one's opponent and by responding to this by showing physical and psychological self-confidence. If such characteristics are included in the definition of the game of chess, present computers cannot play that game, and the question of whether they are better than human beings cannot even arise.

Thus, in order to establish the instrumental effectiveness of computer chess, one must first be able to redefine what it means to play chess. More precisely, it is only the choice of a particular means for testing this effectiveness that requires this redefinition, even if, *after* we have accepted the means in question as adequate, the goal will have been redefined, too. An important consequence of the realization character of science and technology is that the introduction of new "definitions" is much more than a matter of mere terminology. Not just the meaning of chess playing but also its practice has to be transformed. In the present case, the commercial interests of computer firms, the financial interests of the

FIDE (the world chess organization) and of other tournament organizers, and the professional interests of at least some chess players have coproduced this transformation. For instance, computer firms have been willing to sponsor chess tournaments, but only on the condition that some computers be allowed to compete with human players. Then, many organizers and professional players have complied with such arrangements, at least in part for the reason that they were in need of the money (cf. also Nunn, 1990). Furthermore, just recently the FIDE has made a number of basic decisions that imply acknowledging computer programs as legitimate members. Again, money has played a significant role in these decisions. One of the rules accepted says that "humans participating in tournaments with computers will not be allowed to refuse to play against the computers" (see Levy, 1992, 150). So, whether the chess players like it or not, if they want to remain chess players, they are now forced to comply to the new definition of what "playing chess" means. Now and only now, the question of whether computers are better chess players than human beings, and consequently whether computers are effective chess players, can be answered in a "neutral" way, on the basis of "purely empirical evidence."

I have chosen the chess example because it is both simple and illuminating. Mostly, however, the stakes in the realization of science or technology are much higher, as is illustrated by the case of the production of nuclear energy discussed in chapter 6 and by the cases of agricultural biotechnologies presented in chapter 7. My conclusion is that normative philosophy should go beyond the scientistically truncated normativity of normative naturalism in critically analyzing and assessing the claimed effectiveness and reliability of scientifically or technologically realized "means." Put differently, in avoiding the horns of the a priori justificationism of the classical philosophers and the radical descriptivism of the historicist naturalists, we had better opt for an engaged normative pragmatism instead of a quasi-neutral instrumental rationality. Such a pragmatic normativity will be rooted in the practical, historical, and social situation of the philosophers in question. Thus, we are led in a natural way from the question of the normativity of philosophy to the issue of the implications of philosophy's situatedness with respect to its approaches, views, and significance. In other words, this brings us to our final topic, the reflexivity of philosophy.

8.4 PHILOSOPHY AS REFLEXIVE

Generally speaking, reflexive philosophy investigates the "conditions of the possibility" of (scientific, technological, or philosophical) activities and views. Self-referential reflexivity—that is, thinking systematically about the status of one's own claims in the light of their content—is a special case of reflexive philosophy. Traditionally, philosophical reflexivity has taken two opposing courses: a foundational and a skeptical one (cf. Van Woudenberg, 1990). Foundationalist reflexion aims at ultimate foundations of the methods and claims of science, technology, or philosophy. Thus, the universal characteristics of transcendental consciousness, the laws of historical development, or the necessary structure of linguistic communication would provide a criterion for the ultimate justification of human validity claims. Skeptical reflexivity, in contrast, strongly emphasizes the historicity and finiteness of all human achievements. From this perspective, changeable and constructed mental habits, social power, or cultural conventions are taken to be at the root of all validity claims. Hence, philosophical reflexion does not result in a foundation but rather in a deconstruction of any claimed validity of the activities and results in science, technology, or philosophy.

From my point of view, a first requirement of an adequate philosophical reflexivity is that it goes beyond the unfruitful opposition between foundationalism and skepticism. In the end, the search for certainty and the accompanying fear of contingency, which characterize both the foundational and the skeptical version of traditional reflexivity, engender no more than paralyzing frustration or uninteresting philosophy. On the one hand, after more than twenty-five centuries of philosophizing, it should be clear that all attempts at "ultimate foundation" are doomed to fail.[18] Besides, even assuming—for the sake of argument—that universal agreement could be reached on the fact that some claims (for instance, "some propositions are true") are indubitable, this would constitute only a minor step towards the foundation of particular methods or claims in science, technology, or philosophy. The reason is that such "certainties" either are so general as to be almost empty and therefore unsuited for serving as a foundation upon which one can erect any substantial, interesting building; or, if they are made more specific, they will become controversial and thus lose their certainty due to different interpretations of their exact meaning. Thus, the presumed prima facie consensus about the claim that "some propositions are true" will no doubt evaporate fastly in a critical discussion of what is meant by the terms "proposition" and "true." On the other hand, a thorough-going skepticism will not do either. Woolgar's radical program of constitutive reflexivity is a good representative of the skeptical approach. As I have shown in section 5.3, his program proves to be inadequate in several important respects. In the current context, the main point is that the notion of "complete deconstruction" entails a performative contradiction: since any deconstruction presupposes some constructive claims, the program of constitutive reflexivity cannot be consistently realized. Consequently, the goal of a systematic, voluntaristic skepticism is as illusory as the search for ultimate and certain foundations.

What, then, *is* a feasible goal for a reflexive philosophy? It is, I think, no more and no less than taking seriously the fact that the philosophical accounts of the practice of technoscience are themselves part of that very practice. Thus, the aim of a reflexive approach is not to provide an ultimate foundation or a consistent deconstruction but to situate the (theoretical and normative) arguments and views in their historical and social contexts.[19] I will call this approach, which is modeled after and generalized from the discussions in chapter 5, a *differentially situated reflexivity*. Crucially important to it is the insight that in present-day culture, individual students of science and technology, including philosophers, live their lives at different intersections of a manifold of smaller and bigger contexts. Because of this, they will be differentially situated with respect to theoretical and normative issues taking place in different contexts. In other words, their situation is not a monolithic whole, which would unambiguously determine their approaches, skills, and views. This differential situatedness entails not only the unease of the "death of the unified subject"; because we never fully share our situation with others, it entails as well the necessity of genuine communication and thus constitutes a significant source of creativity and change.[20]

In the context of theoretical and normative discussion and controversy among philosophers from various quarters, this implies that there is not one Great Divide between "us" in our situation and "them" in theirs. Generally speaking, there will be at once many points of agreement and many points of disagreement. In a number of cases, general assumptions, basic skills, and specific views will be widely shared, thus enabling what is called "rational argument." Regarding other questions, matters may be more ambiguous, implying that discussants will try to persuade each other from different perspectives. Finally, broad and substantive disagreements may exist, leading to polemic, struggle, or exclusion.

So far these claims are rather abstract. It is not difficult, though, to specify them through examples taken from the preceding chapters. As a preliminary observation, it is good to keep in mind that scientists are also differentially situated and not caught in monolithic, closed frameworks or paradigms. As we have seen in the discussion of the generalized correspondence principle, in chapter 3, in practice the situation of the relevant scientists displays not just discontinuities but also continuities with other (earlier) situations. And in the case of experimental science, as discussed in chapter 2, reproducibility, in any of its forms, constitutes a common, nonlocal norm across widely divergent situations. With respect to philosophical claims, the preceding chapters contain various illustrations. For example, the claim that we cannot get rid of the ozone hole simply by stopping scientific discourse about it, used in one line of argument in section 5.5, appears to be a widely shared assumption in our culture. The critique of technology as a neutral tool, as put forward in section 6.4, addresses a more disputed area. On the one hand, the arguments as such appear to be quite straightforward and not easily refutable. On the other hand, as we have seen once more in the discussion of normative naturalism in the preceding section, it is also the case that in our "technological culture," the notion of a neutral, instrumental rationality keeps exercising its powerful grip (cf. Gault, 1991). As a result, even if the arguments as such are conceded to be all right, dispute often continues as to their significance. Thus, an important task concerning this issue is to offer convincing analyses and illuminating cases that demonstrate the normative weight of the critiques. Finally, there are philosophical positions with hardly any overlap with the preceding accounts. One example is hard-boiled scientism, the blind faith in the unique truth of scientific knowledge and the unambiguous superiority of a science-based technocracy. At the other extreme we find philosophical views that—in being congenial to forms of religious or ethnocentric fundamentalism—legitimize systematic oppression of those who are "different." In philosophical clashes with such positions, polemic, irony, or neglect appear to be the most adequate responses.

Adding reflexivity to the theoretical and normative aspects of our philosophical studies of science and technology has a further important consequence. It implies that the role of philosophy in wider social controversies regarding science and technology cannot be that of an enlightened vanguard or an impartial outside arbiter, who is able to reveal the final truth about the theoretical and normative issues in question. Being differentially situated implies that our accounts should rather be seen as specific parts of, and we ourselves as specific participants in, the relevant debates. The analyses in chapters 6 and 7 about the social role of (appropriate) technology are meant to illustrate these points.

In sum, a reflexive philosophy that does not hide its own roots will be seen to employ a whole range of discursive procedures, from logical argument to political polemics and from conscious neglect to skillful persuasion. In doing so, it will be at once rational and partial, argumentative and interested, theoretical and normative. Such an approach is feasible because the "frameworks," "practices," "paradigms," "language games," "contexts," "cultures," or "forms of life," in which we live our lives, are both differentiated and nonlocally patterned. Thus, given the differential *situatedness* of our philosophizing, we are not able to take part in a Habermasian, ideal speech situation for the purpose of arriving at a universal, rational consensus. But due to our *differential* situatedness neither are we caught in a Rortyan *ethnos* of a monolithic "us." There is a nonlocal space between the horns of an impossible universalism and a stifling ethnocentrism (cf. Van den Belt, 1991). As the chapters of this book intend to show, in our current situation this space is wide enough for practicing a meaningful theoretical, normative and reflexive philosophy. A philosophy, in other words, that is at once in and about the world.

NOTES

1. INTRODUCTION: REALIZATION AND NONLOCALITY IN SCIENCE AND TECHNOLOGY

1. See Fuller, 1991, 292–293.

2. REPRODUCTION AND NONLOCALITY IN EXPERIMENTAL SCIENCE

1. Thus, my approach is not that of the "testing theories of scientific change" project. See Laudan, Laudan, and Donovan, 1988, and compare the critical remarks in Nickles, 1986.
2. As explained, the philosophical discussions in the present chapter mainly turn on the reproducibility of experiments. In addition to this, the proposed analysis of experimenting immediately suggests several further elaborations. In particular, I think of the relevance of experimenting for the problem of scientific realism; aspects of the relationship between science and technology other than those dealt with here; the similarities and dissimilarities between experimentation and observation; and the problems concerning the nature of experimental testing. Apart from the latter, these issues will be discussed in the chapters 4 and 6.
3. The discussion in this section is partly a summary of and partly a more precise statement and further development of the analyses presented in Radder, 1988, 59–76. The idea of differentiating the theoretical description of an experiment from its material realization was stimulated by Habermas's distinction between theoretical discourse and instrumental or experimental action. See Habermas, 1973, and the postscript of Habermas, 1978.
4. Thus, the general term "object" may stand for a thing, a process, a phenomenon, etc. Janich, 1978, 9–19, discusses various kinds of apparatuses and their functions.
5. For more details of this example, see Radder, 1988, 59–69.
6. For this notion of closedness, see Radder, 1988, 63–69, and also chapter 6.
7. Already in the 1920s and 1930s the importance of experimenting as concrete action was stressed by Dingler. However, in his wish to ground natural science in practical action by means of operational definitions, he understated the significance of theoretical knowledge, including its importance in carrying out experiments. See Dingler, 1952, for a concise statement of his views on the matter. For a recent Dinglerian view, see Tetens, 1987. (Remarkably enough, Dingler and his philosophical ideas were intimately connected with the attempts to

build up a "German physics," a physics that should agree with the *Blut-und-Boden* doctrines of National Socialist ideology. See Richter, 1980.)

8. Note that the above definition neither fixes the nature of this common language nor claims any ontological privilege for it. In practice, it may have various characteristics, as long as it enables us to materially realize experiments in the way specified above. For philosophical purposes this is a crucial point: see Radder, 1988, 76 and 105–106.

9. See the first of the two illustrations given in this section.

10. Note that this account does not make use of the assumption that the theoretical interpretation of experiments can be done in terms of stable, low-level theoretical "home truths" and therefore does not depend on the ever changing high-level theories (see Hacking, 1983b, e.g., 265). On this issue I agree with the criticism put forward by Morrison, 1990, 6–14. In Radder, 1988, 91–93 and 144–147, I have presented counterexamples to the claim that home truths remain stable across time. On the other hand, the above analysis of experimenting seems to be more in line with an aspect of Hacking's position that is emphasized by Stump, 1988, namely, that skillful experimentation is relatively autonomous because it need not be committed to one specific theoretical interpretation.

11. This study has been reported extensively in Lynch, Livingston, and Garfinkel, 1983. The quotation is from p. 225.

12. See Popper, 1965, 45–46; Habermas, 1978, esp. ch. 6; Bhaskar, 1978, esp. ch. 2. Other terms used are "repeatability," "replicability," and "regularity."

13. Since reproducing experiments is an important way of making them "public," the fact that achieving the reproducibility of the material realization of an experiment is more difficult in an experiment's "exploratory" or "trial" stage than in its "demonstrative" or "showing" stage, is of course precisely what we would expect. For these distinct stages of experimentation, see Gooding, 1985, and Shapin, 1988, 399–404.

14. The above argument will be discussed in detail in chapter 4. Cf. also Giere, 1988, 108–109.

15. Occasionally I will use the less accurate phrase "reproducibility of the theoretical description" or of the "result" as a shorthand for the reproducibility of the experimental situations to which these descriptions refer.

16. Of course, for more particular purposes, further specifications may be useful and clarifying. See, for instance, the taxonomy of the elements of laboratory experimentation offered in Hacking, 1988a, 508–511.

17. Cf. Collins's distinction between algorithmic and enculturational models of experimentation (Collins, 1975, 206–208). See also Ravetz, 1973, ch. 3.

18. Here lies an interesting parallel to the individual bodily "sensibility" of human beings and the problems this poses for an exclusively natural scientific approach to medicine and health care. See Kunneman and Hullegie, 1989, for an illuminating account of these problems.

19. Collins, 1975, 211. See also Rouse, 1987, 86–92.

20. See Salmon, 1984, 213–217, and Cartwright, 1983, 82–85 for brief accounts of some of these replications.

21. Shapin and Schaffer, 1985, 43. I will discuss the role of witnesses in section 2.7.

22. Shapin and Schaffer, 1985, 43. Thus, this experiment exemplifies the fact that reproducing an experiment under a fixed theoretical description is compatible with slightly different material procedures (a glass globe of a somewhat different size had to be mounted upon the pumping device). That is, in this case a reproduction of the experiment under a fixed theoretical interpretation implies an *approximate* reproduction of its material realization.

23. For another example see Collins, 1985, 67–68.

24. Cf. Hacking, 1983b, 22–24. The reproductions of a TEA-laser, discussed in Collins, 1985, ch. 3, are also of this kind; the scientists in question reproduced this device in order to use it in other experimental work. Calibration techniques that, if successful, show the ability of parts of the experimental setup to reproduce known results provide another illustration. See Franklin, 1986, 175–181, and Hones, 1990.

25. On this specific issue I agree with Collins, 1985, 14, who states that "there must be something more to a rule than its specifiability." Cf. also the discussion of "values, norms and practice" by Henderson, 1990, 128–133. For instance: "Much of what we learn is socially taught; yet once learned, it informs and constrains social practice. Thus, practice and value should not be set apart as largely distinct matters" (p. 128).

26. Please note that I do not say that a reproducible material realization is sufficient for accepting and realizing the experiment (or technology) in question. Other (social, ethical) considerations are also important in this respect. See chapter 6, sections 4 and 5.

27. Constitutive norms or rules (e.g., of a game such as chess) are definitory: if we do not follow them, we simply are not playing the game (correctly). Regulative rules are like the strategic or tactical rules of a game: on the average it helps to follow them, but they cannot guarantee success in a particular case.

28. This conclusion agrees with Galison's account of experimentation, especially with his views on experimental constraints and experimental stability (see Galison, 1987, ch. 5). For more general discussions of the significance of nonlocal patterns in scientific practice, see chapter 4 and section 5.4.

29. In this sense I disagree with Rouse's claim that even where stabilization or (to use his term) standardization obtains, experimental knowledge remains essentially local knowledge. See Rouse, 1987, 111–119.

30. Such claims, or comparable ones, have been rather influential in recent science studies. For other case studies and/or further discussion on the issue, see Schaffer, 1989; Draaisma, 1989; Brown, 1989, 76–86; and Gooding, 1990, ch. 8.

31. My discussion is primarily based on Collins, 1985. In this book Collins summarizes, systematizes, and develops much of his earlier studies. In Collins, 1990, he appears to qualify his views in a certain sense. For instance, he now admits that "the idea of tacit, inexpressible knowledge doesn't leave enough space for the fact that more and more detailed descriptions of actions can be extracted from us" (pp. 93–94). However, he also claims that this fact does not lead to any new philosophical conclusions (pp. 47 and 110–111), and accordingly his account of the experimenters' regress (pp. 184–186) remains unchanged.

32. This topic is discussed more extensively in chapter 4.

33. Brown, 1989, 86–93, makes a number of clarifying comments on the issues of tacit knowledge and the experimenters' regress, which are to a certain extent in agreement with the point made above. Yet, in his discussion he wrongly jumps from the relativist extreme to its contrary, the rationalist extreme. Brown does acknowledge the significance of skillful experimental action (p. 87), but he appears to shrink away from the consequences of this fact. He first separates the issue of tacit knowledge from the issue of the experimenters' regress by focussing on what I have called the "first formulation" of the regress. On this premise, he then claims that background theories "can be employed to block the regress" (p. 91). This claim, however, straightforwardly contradicts the plausible part of Collins's views: explicit, theoretical knowledge is in general *not sufficient* for competently materially realizing experiments. Brown seems to sense the difficulty but, instead of argumentatively substantiating his views, in the end he relapses into the familiar but unconvincing "in principle" strategy: "We have a large number of background beliefs which will tell us what gravity waves are and how they can, *in principle*, be detected" (p. 90, emphasis added).

34. Note that they could retain their philosophical claim if they weaken it somewhat to the effect that "systematic scientific theories aim at explaining nonobservable phenomena and not data, whether the latter are observable or not."

35. For a discussion of whether or not this claim is plausible, see chapter 4.

36. Shapin stresses the notion of participation in the context of the Royal Society: "The Royal Society expected those in attendance to validate experimental knowledge as participants, by giving witness to matters of fact, rather than play the role of passive spectators to the doings of others" (1988, 390).

37. The point is that experimental natural science often is one of the factors contributing to a successful technology. It is by no means implied that the process of transforming scientific experiments into technological artifacts is a simple or unidirectional matter of "application." See Latour, 1983; and also chapters 6 and 7.

38. Please note that these considerations refer to de facto functioning legitimations and aim at elucidating these somewhat further. I do not think that these legitimations are satisfactory. In fact, I even think that the entire enterprise of justifying experimental natural science "as such" or "as a whole" (in contrast to specific developments, results, or uses) is mistaken. A systematic argument for this is that, although experimental science does exhibit nonlocal patterns (see section 2.4), it is equally the case that its concrete development, results, and uses depend upon local characteristics. As a consequence there cannot be a general justification of science as such. See also Radder, 1988, 118–119, for my objections to Habermas's notion of a "technical interest" as being generally constitutive and justificative of the natural sciences.

39. Of course, there might be other variances between these "materially equivalent" interpretations that may make some difference with respect to the stability of the relevant experimental technologies.

40. Latour notes this meaning of "black box" (1987b, 2–3), but further on in his book, he changes it in favor of the definition given above.

3. HEURISTICS, CORRESPONDENCE, AND NONLOCALITY IN THEORETICAL SCIENCE

1. See Radder, 1988. Cf. also Hakfoort, 1986, ch. VI, who argues that in addition to Kuhn's mathematical and experimental traditions, the tradition of "nature philosophy" is essential to the development of science.
2. This is completely analogous to Krajewski's approach. Krajewski does not speak of meta-argumentations but of an "intermediary product of reasoning" (1977, 49).
3. For a more extended account of this case see Radder, 1988, 124–144.
4. This point has been insufficiently noticed in the historiography of quantum mechanics so far. See Radder, 1988, section 5.2, for more details.
5. Letter from Bohr to Rutherford, 27 December 1917, reprinted in Bohr, 1976, 682–683. Compare also Epstein's reaction to Bohr's 1918 paper "On the Quantum Theory of Line Spectra":

> it really seems to me that the difference between a quantum mechanical and a classical treatment is not at all as large as we assumed up to now. (letter from Epstein to Bohr, 14 May 1918, reprinted in Bohr, 1976, 637)

And, in another context, Sommerfeld wrote:

> Hence one is led to the gratifying point of view that the contradiction between mechanics and quantum theory is not as sharp as it seemed to be up to now: when we correctly apply the quantum theory . . . the basic principles of mechanics remain valid for molecular transitions from the initial to the final state. (Sommerfeld, 1917, 502)

6. Of course, this does not imply that all differences between the theories have vanished. The stability of stationary states and their discreteness remain classically inconceivable.
7. Cf. Hendry, 1982; Radder, 1982. Nevertheless Bohr as well as Kramers retained a measure of conceptual correspondence in their interpretation of *radiation*. On the basis of the assumption that a quantum theory of radiation should be a "natural generalization" of the electromagnetic wave theory, they strongly opposed the light quantum hypothesis. In 1925, however, the conceptual correspondence for radiation proved to be untenable as well.
8. Of course, from a physical point of view putting $h = 0$ is impossible: Planck's constant h is an empirically fixed *constant* unequal to zero (cf. also Krajewski, 1977, 9–11). What we mean physically is that for classically described macroscopic bodies, h is negligibly small as compared to the macroscopic actions. As we will see below, it is possible to work this out in detail for specific domains with a view to establishing (approximate) numerical correspondence. In formal correspondence, which is at issue here, these physical arguments are of little help. In this case we can only obtain the classical counterparts of quantum theoretical equations by simply putting $h = 0$.
9. An ensemble interpretation of the quantum mechanical state function will not be able to bridge this gap either. Apart from the specific problems of this

interpretation, fundamental differences will remain between classical and quantum ensembles. See, e.g., Park, 1968, 212–217.

10. Laudan, 1981, 39–40. Please note that this point does not invalidate Laudan's criticism of convergent realism as such (see also section 3.7).

11. In order to be complete, we should add here a discussion of the *accuracy* of the numerical values. For this, see Fadner, 1985, 832–835.

12. I leave aside the question of correspondence in nonmathematicized sciences for further study. Is there no correspondence in these cases, or does a functional equivalent of formal-mathematical relations between theories exist? Cf. Koertge, 1969, for relevant case studies.

13. Other examples can be found in Darrigol, 1986. He writes:

> these examples show that permanent formal schemes allow transfers of knowledge between successive theories even if their basic concepts appear to be incommensurable, even if their inventors' world views conflict. (pp. 198–199)

14. See, e.g., Shapin, 1982; Knorr-Cetina and Mulkay, 1983; Latour, 1987b. Note that the above argument refers expressly to conceptual aspects. Therefore it should not be taken to imply that facts are *no more* than social constructions (see also section 3.7 and chapter 4). However, the arguments of the present chapter should not be read either as saying that the mathematical and experimental aspects of natural science are not amenable to sociological analysis, because these aspects would always be universally agreed upon. Instead the claim is that there are some philosophically relevant ways in which particular mathematical and experimental nonlocalities make it possible to bridge differences between "local scientific cultures."

15. Zahar is somewhat ambiguous on this point. At one place in his 1983 he states that heuristics "operates purely deductively" (p. 245). But elsewhere he asserts that the logic of discovery is only "largely deductive" (p. 249) and that there may be small logical gaps that, however, can be bridged by "considerations of simplicity or convenience" (p. 252).

16. Compare also Goodman's green-grue paradox and the conclusions he draws from this "new riddle of induction" (Goodman, 1973).

17. For a more extended review of this methodology, see Kirschenmann, 1985, 1990.

18. Cf. Latour and Woolgar, 1979, 168–174. Of course, something is going on in the heads of discoverers, but since the same holds for someone proposing a certain justification, this fact by itself does not differentiate between discovery and justification.

19. Hoyningen-Huene, 1987, rightly points out that the contrast between factual and normative claims forms the core of the discovery-justification distinction. Therefore the above formulation is certainly not intended as an exhaustive definition of the business of philosophy of science. A further reason for abandoning the discovery-justification distinction is that "justification" is a much too narrow specification of philosophy of science's normative and critical task: tied as it is to a predominantly nonsocial epistemology, it tends to neglect or take for granted the role of the sciences as social practices. In chapter 8 I will deal with the issue of the status and role of philosophy (of science) in a more systematic manner.

4. SCIENCE, REALIZATION, AND REALITY

1. See, e.g., Kuhn, 1970b; Knorr-Cetina and Mulkay, 1983. For a recent statement of the challenge, see Feyerabend, 1989.

2. See Bachelard, 1984 (originally, 1934), 1972. Compare also the discussion of this aspect of his work in Latour and Woolgar, 1979, 63–69.

3. If we have to deal with experiments on human beings, the notion of independence has to be slightly qualified to read: independence of the existence of the experimenters. From now on, every time I speak of a "human-independent reality," I will take this qualification for granted.

4. See, e.g., Putnam, 1975, the papers "Explanation and Reference" and "The Meaning of 'Meaning.'" Cf. Radder, 1988, esp. 109–115.

5. For the notion of instrumental realism, see Baird, 1988, and Ihde, 1991.

6. Only in section 4.6 will I briefly discuss two alternative responses to the Bachelardian challenge.

7. See for some highlights Latour and Woolgar, 1979; Hacking, 1983b; Collins, 1985; Shapin and Schaffer, 1985; Franklin, 1986; Galison, 1987; Gooding, 1990. For a review, see Hacking, 1989b.

8. See chapter 2. In the present section I speak of "reproducibility" in an undifferentiated manner, because my aim here is to set out the theoretical-philosophical argument as clearly as possible. In fact, as we have seen, reproducibility is a complex notion, and it is necessary to differentiate between different types and ranges of reproducibility. In the next section, I will be more specific in this respect.

9. Distinguishing reproducible experiments from their realizations was stimulated by, but is nevertheless essentially different from, Bogen and Woodward's distinction between "phenomena" and "data." See Bogen and Woodward, 1988, 305–322, and compare my discussion of their views in section 2.6.

10. In Radder, 1988, esp. 115–117, I dealt with this issue in terms of the idea of stable "aspects" of reality. The present discussion aims at clarifying and developing this earlier approach.

11. The epistemological criterion of (co)reference summarized in section 4.2 tells us how we can come to know whether or not the term in question refers.

12. Cf. on this point Hacking, 1988b, esp. 284–286.

13. An important question that arises from this discussion concerns the status of the nonreproducible, or the unique. My present feeling is that the nonreproducible, in contrast to the nonreproduced, is in principle unknowable. This is, however, a complex issue, which I cannot enter into here.

14. Thus, the present notion of realization includes, but is also much broader than, the notion of material realization discussed in chapter 2.

15. See Pickering, 1989, 1990, concerning coherence; and Gooding, 1990, ch. 7, concerning convergence. From my point of view these "negative realisms" may be interpreted as limiting cases of referential realism, obtaining for rather restricted ranges of reproducibility.

16. Therefore, Collins's coupling of the notions of replication and induction is, I think, less fortunate. Compare the subtitle of his 1985 book, *Changing Order: Replication and Induction in Scientific Practice*.

17. Of course *q*, as a concrete experimental result, cannot be obtained independently of any context whatsoever.

18. Note, however, that this view by no means entails a commitment to essentialism, in the sense that the referring of a replicable experimental claim would provide a good reason for interpreting the claim as revealing the immutable essence of the relevant elements of reality. The relativization from "universals" to "nonlocals" excludes such an interpretation. Thus, the above view is consistent with the plausible assumption that classification and concept formation is always relative to specific theoretical perspectives, social interests, or human goals.

19. Locke, quoted in Blokhuis, 1985, 35.

20. See, e.g., the views of Cassirer (as summarized in Blokhuis, 1985, 175–179); Popper, 1965, 64–68 and 420–426; Koningsveld, 1973, esp. ch. I.

21. Thus, for this reason Rouse's criticism of Heidegger's early philosophy of science is not fully convincing. See Rouse, 1987, 73–80.

22. See Keller, 1985, esp. 129–135. See also Cartwright, 1983, who shows convincingly that, due to the fundamental significance of "unjustifiable" but "working" experimental and theoretical skills and methods, the success of experimental realizations can by no means be attributed solely or primarily to the validity of the laws of nature.

23. Compare the alternative account of closed, experimental and technological systems in Radder, 1988, 63–69; see also section 6.2.

24. The controversy between Boyle and Hobbes (see Shapin and Schaffer, 1985) illustrates the essential contestability of the experimental method.

25. Thus Collins, 1985, 127, rightly remarks: "The fact remains that our experience of nearly all natural phenomena is like the experience of laser building: we know that the familiar objects of science are replicable."

26. In the next chapter, I will offer a more detailed discussion of constructivist forms of localism and associated types of relativism.

27. Pinch, 1985, 19. Pinch remarks that the account has already been greatly simplified. The other case described in his article, the observation of solar neutrinos, is also illuminating.

28. Note that this is a theoretical definition. It does, of course, not claim to represent the usage of the terms "observation" and "experiment" in scientific practice.

29. See Hacking, 1983b, 263. Compare also the previous quotation.

30. See, especially, Morrison, 1990. Note that the above account of experimentation, and hence referential realism, does not depend on a distinction between experimenting with and experimenting on entities. Instead, it is based on an analysis of the complete material realization and theoretical interpretation of the overall experimental processes (see also section 2.2).

5. NORMATIVE REFLEXIONS ON CONSTRUCTIVIST APPROACHES TO SCIENCE AND TECHNOLOGY

1. The point has also been made by other authors: see, for example, Chubin and Restivo, 1983; Russell, 1986; Horstman, 1988; Fuller, 1988, 266–268; Pels, 1990, 19–20; Lynch and Fuhrman, 1991.

2. The Strong Programme will be largely left aside. For a critique of its neglect of normative questions, see Lynch and Fuhrman, 1991.

3. For some examples see the Abstracts of the 1988 Joint 4S/EASST Conference, 1988, the papers by Brouwer (pp. 101–102), Oudshoorn (pp. 163–164), Potter (p. 171), and Van den Wijngaard (pp. 202–203). Despite these examples, Keller is right in claiming that sociologists of scientific knowledge have been largely gender blind: see Keller, 1988.

4. And for good reasons, it seems, since, as Barnes (1981, 492) once subtly remarked: "With cream-cakes there is a chance of satisfying hunger—with accounts of cream-cakes there is not."

5. Such a "let a hundred flowers bloom" strategy seems to be advocated in Latour, 1988, 169–175.

6. Even Latour complains: "if only God—or Mammon—willing, I could write in my own mother tongue!" (1988, 171).

7. Some constructivists hold a more balanced view, in which both change and continuity and both actors and structures have a place: see especially, Bijker, 1993. It seems to me, however, that conceding this much takes you well beyond constructivism (cf. also Hagendijk, 1990).

8. In a similar vein, Nickles, 1988, argues for the methodological significance of reconstruction in the practice of science. His distinction between "single-pass" and "multipass" accounts of science is especially important in this respect. In section 8.2 I will return to this topic.

9. See also chapter 3. For other cases, see Galison, 1987, esp. 10–13; Downey, 1988; Star and Griesemer, 1989.

Also, Wynne acknowledges the existence of more general patterns: "The technology should embody some general principles (it should not work only by chance) because otherwise there is no reproducibility, or transportability to other contexts" (Wynne, 1988, 152); and: "Thus although rationalization of social worlds takes place to meet the 'imperatives' of a (universal) technology, at the same time differentiation of the technology takes place in order to correspond with the variety of social and physical worlds in which it is enacted" (Wynne, 1988, 154). Unfortunately, perhaps as a consequence of his view that "rules follow practices," we read in Wynne's paper far less about the role played by these "general principles" and "rationalization processes" than about the "unruly," local practices.

10. For this notion, see Fine, 1986, chs. 7 and 8; cf. also Rouse, 1987, ch. 5.

11. For example, writing and acknowledging the history of minority groups from their own perspective is widely considered to be important with regard to the possible realization of their normative goals. I will come back to this example below, in the discussion of the actor-network theory.

12. More generally, epistemological relativists hold that knowledge claims cannot be divided unambiguously into rational and irrational ones, where "rationality" may also be explained in terms of adherence to a justified methodology rather than in terms of a correspondence to reality.

13. Woolgar, 1988a, 73. Similarly: "a major thrust of post-modern critiques of science is to suggest the essential equivalence of ontology and epistemology: how we know *is* what exists" (p. 54).

14. Since the term "discourse" is meant to include all the relevant research practices of the scientists, "stop discoursing" implies and requires more than just "stop talking."

15. Comparable issues can be found in the area of medicine and illness. Although I agree that "iatrogenesis" is a real phenomenon, this does not imply that all diseases are no more than the product of medical practice. In analogy to the case of the hole in the ozone layer, one could pose this question: how many constructivists would be prepared to claim that the "problem" of lung cancer among heavy smokers can be solved by stopping medical discourse about it, *even if* the patients in question continue to smoke their forty cigarettes a day?

16. Collins, 1985, 16 (emphasis added). Also Knorr-Cetina and Mulkay, 1983, 5–6, point to the general significance of this type of relativism to social studies of science.

17. See Radder, 1988, esp. ch. 4, and the summary in section 4.2 in this book. So, I agree with Woolgar that the conceptual representations of science should not be interpreted in a realist manner. But in contrast to Woolgar, I do not think that, therefore, the whole practice of representation should be revolutionized. As I argued in the preceding section, in specific contexts specific representations may well be plausible, reliable, useful, or desirable.

It will also be clear that my referential realism is more modest than the transcendental realism advocated by Bhaskar (see section 4.6). In particular, I do not claim that scientific realism is *required* for human emancipation (cf. Bhaskar, 1986).

18. The arguments in this section aim at a *reductio ad absurdum* of ontological and epistemological relativism. Note that the argument does not work for all forms of antirealism. It does not, for instance, apply to Van Fraassen's constructive empiricism: see Van Fraassen, 1980. His position, however, is implausible as a consequence of other intrinsic problems: see Radder, 1989, esp. 300–304.

19. To adapt an example of Isaiah Berlin's.

20. Some authors claim that the actor-network theory may be conveniently combined with a systems approach to technology (e.g., Law, 1987, 112–114; MacKenzie, 1987, 196–199; for the systems approach, see Hughes, 1987). Such a combination may be accomplished if attention is also paid to the system builders (not only to the system itself), if the boundary between system and environment is seen as flexible, and if the occurrence of conflicts, alternatives, and the like is taken into account.

21. In this section, I make use of the "winners-losers" terminology, which is suggested quite strongly by the military metaphors of the actor-network theory, only for the sake of argument. I do not mean to imply that all developments in science and technology can be adequately characterized in terms of winners and losers.

22. See Law, 1987. Latour's notion of "cycles of accumulation" is also developed from the perspective of the Princes' centers of power (Latour, 1987b, 219–223); the perspective of the domesticated actors is strikingly absent from these pages.

23. See also the next chapter, where I will discuss the notion of a "closed system" in detail.

24. In this respect my approach differs considerably from that of some other, recent philosophical critics of constructivism, such as Slezak, 1989, or Roth and Barrett, 1990.

6. EXPERIMENT, TECHNOLOGY, AND THE INTRINSIC CONNECTION BETWEEN KNOWLEDGE AND POWER

1. In doing so, I develop and generalize notions introduced and discussed in Radder, 1988, 59–76 and 174–175.

2. In the past decade there has been a growing interest in technology on the part of historians, sociologists, and philosophers (see, e.g., Laudan, 1984; Pinch and Bijker, 1984; Sass, 1984; Krohn and Weyer, 1989). In many cases, the studies in question approach technology by looking for analogies or for similarities and differences between science and technology. Thus, Pinch and Bijker, 1984, and MacKenzie, 1989, try to export methods and insights from the sociology of scientific knowledge to the field of technology studies; Constant, 1984, argues that Kuhn's model of scientific paradigms and scientific communities can be fruitfully applied to the development of technology; and Latour, 1983, even claims that, in crucial respects, there is no "distinction" or "separation" between "inside and outside the laboratory": the only differences are differences of scale.

3. A fine example of the issues under discussion can be found in Levi, 1986, 204–210. He tells "a story of silver" that strikingly illustrates the sociocognitive complexities of maintaining closedness during the technological production of X-ray photography paper.

4. Given the qualification of closedness as theory laden, in this chapter the notion of reproducibility should be taken primarily as reproducibility under a fixed theoretical description (see chapter 2).

5. This implies that such a distinction between experimental facts and technological artifacts as is employed by Pinch and Bijker, among others, is not adequate: see Pinch and Bijker, 1984, e.g., 424. Just as in technology, one strives in science for the closedness of systems, not only for the closure of debate. Thus, in this respect my analogy between science and technology is stronger than the one proposed by Pinch and Bijker.

6. Regarding the nuclear energy case, I make use of two reports in particular, both written by the Stuurgroep Maatschappelijke Discussie Energiebeleid (the "Steering Committee" of the BMD, the Dutch debate on energy policy). The reports are referred to as Stuurgroep, 1983, and Stuurgroep, 1984. On the present point of social conditions, see Stuurgroep, 1983, 97–98, and 1984, 210.

7. This approach to the above problems seems to be congenial to ideas propounded in Wynne, 1983. However, the precise meaning of Wynne's main concepts ("social enactment" and "social viability" of technology), and especially their relationship to the cognitive aspects of technology, remains somewhat unclear.

8. These examples clearly refute Constant's claim of the *objectivity* of technological failure. He states that "all functional-failure based problems or anomalies are objective; a specific system really does not work very well in a specific context" (Constant, 1984, 31).

9. Again, I have to add that the above analysis in terms of closedness bears on a specific aspect of experimental and technological systems. Apart from this, problems may also arise *within* these systems, regardless of their interaction with the environment. One type of problem that is especially important in technology (though it plays a role in experimentation, too) stems from the occurrence of so-called conjunctures (Bhaskar) or combined causes (Cartwright) or interference effects (Hacking). That is to say, it stems from situations in which several different mechanisms simultaneously produce significant effects. The problems then arise because, in many cases, we have no, or no adequate, knowledge of what will happen when the relevant mechanisms operate in combination, even if the effects of the separate mechanisms are well known. Cf. Bhaskar, 1978, esp. ch. 2; Cartwright, 1983, esp. chs. 2 and 3; Hacking, 1983a, 1986.

10. See also the criticism of Collins's experimenters' regress in section 2.5, the "parable of the extraterrestrials" at the end of section 5.5, and the comment on Pinch and Bijker's approach at the beginning of the present section.

11. With respect to the problem under discussion, the views of Callon and Law, 1982, and also Latour's later, 1987b, views, which take into account the interplay of processes of enrollment *and* counterenrollment within networks of actors, are more differentiated.

12. To indicate the size of the BMD: its total budget was HFl. 28 millions.

13. Due to the accident at Chernobyl, in April 1986, these plans did not materialize.

14. Cf. also Wynne's, 1983, criticism of the "tool" concept of technology.

15. Stuurgroep, 1984, 212 (and 224). I have to add that at another place (p. 254), the report is more careful regarding the possibility and value of quantitative risk analysis.

16. Thus, Habermas's distinction between (nonsocial) instrumental action and (social) communicative action is inadequate: see Habermas, 1970, 1978.

17. I do not claim that this is the only intrinsic connection between knowledge and power. See also the analyses by the Radical Science Journal Collective, 1981, 13–20, and by Rouse, 1987, esp. ch. 7.

18. Cf. Keulartz, Kwa, and Radder, 1985. In this paper we have analyzed the controversy between regular and alternative medicine along the above-mentioned lines.

19. Concerning alternative possibilities and realities, cf. also Noble, 1979, esp. 45–50.

7. THE APPROPRIATE REALIZATION OF TECHNOLOGY: THE CASE OF AGRICULTURAL BIOTECHNOLOGY

1. As already mentioned in the *Preface*, the present chapter is a slightly adapted version of a paper written together with Joske Bunders. For this reason I have retained the plural "we" throughout the text.

2. See Gotsch and Rieder, 1989; Röbbelen, 1990; Fulkerson, 1991; Van de Bulk, 1991; Duesing, 1992; Hillman, 1992; Jones, 1992; Lamb, Ryals, Ward, and

Dixon, 1992; Lindsey, 1992; Scovel, 1992; Beck and Ulrich, 1993; Fraley and Schell, 1993; Somerville, 1993.

3. To avoid misunderstanding, we want to stress that our present discussion is not meant to imply that all normative guidelines are pointless anyway. As will become clear later, we think that a sensible approach is possible when one takes into account the significant differences between the various biotechnologies and when one analyzes the particular technologies in question from a more comprehensive perspective.

4. In this discussion we have made a fruitful but eclectic use of Bijker, Hughes, and Pinch, 1987, especially the chapters by Hughes, Callon, Law, and Constant. Note that the key features are not meant to supply a rigorous *definition* of technology.

5. Compare the discussion of the relationship between actor-networks and technological systems at the beginning of section 5.6. Note also that—in line with my pragmatic usage of the systems terminology—I here employ the notion of technological system in a somewhat wider sense than in the preceding chapter. The more specific discussion of closedness requires a distinction between system and environment. The present approach stresses the significance of systemic interaction, coordination, or integration more generally, also between closed "systems" and their material and social conditions. Thus, the analyses from the preceding chapter should be seen as pertaining to a set of specific issues within the framework of the present chapter.

6. Note that this definition differs somewhat from the one proposed in Bunders, 1990, 81.

7. The biotic elements in the analyses later in this section will mostly be plants and microorganisms. Nevertheless, after a suitable qualification of the notions of "material," "psychological," "social," and "cultural," the framework is, we think, applicable to the case of animals as well.

8. The problems mentioned might be dealt with by means of chemical, biotechnological, or biological techniques. This would, however, not automatically enhance the appropriateness of the overall technology. Chemicals, for example, will generally be too expensive for the large majority of the small-scale farmers.

9. This approach has been developed by the research team of the Department of Biology and Society of the Vrije Universiteit, Amsterdam. See Bunders, Broerse, and Stolp, 1989; Bunders, 1990; Bunders and Broerse, 1991; Brouwer, Stokhof, and Bunders, 1992.

8. PHILOSOPHY: IN AND ABOUT THE WORLD

1. The discussions in this book focus on the philosophy of science and technology. It would be interesting to examine the implications of the proposed approach for other areas of philosophy. An attractive attempt in the history of philosophy is Stone's, 1988, account of the trial of Socrates, which clearly contains theoretical, normative, and reflexive elements.

2. Thus, Gooding's criticism of the notion of material realization misses the point (see Gooding, 1990, 213). Since this notion is primarily theoretical, it does

not aim at describing the "observable learning processes of individual scientists." Furthermore, it is of course evidently true that in processes of division of labor—which are becoming increasingly important in modern science—the people involved acquire a less complete knowledge and skill than in cases where they do both the theoretical and the material work.

3. Lynch, Livingston, and Garfinkel, 1983, 206 (emphasis added). See also, Lynch, 1992; and compare the discussion in Woolgar, 1988a, ch. 6.

4. For a detailed review, see Suppe, 1977, 1–118 and 617–632.

5. See Van Fraassen, 1980. Note that Van Fraassen's empiricism concerns science and not his own account of it.

6. See for example the arguments in section 5.5 concerning the issue of ontological, epistemological, and methodological relativism in constructivist approaches to science and technology. There I argued that even the least assuming position, methodological relativism, needs a philosophical underpinning if it is to have the broader significance that its advocates attribute to it.

7. Cf. the critiques by Russell, 1986, 334–335, and by Winner, 1993, 440–443.

8. Remarkably enough, Bijker's meta-account displays some notable agreements with what was once called the "received view" in the philosophy of science, a view that also resulted from a watering down of a more radical empiricist conception. See, e.g., Nagel, 1961.

9. In contrast to Laudan, 1993, I think that, in spite of the presence of some obstacles, there are interesting opportunities for a fruitful interaction between philosophy of science and technology on the one hand and present-day science and technology studies on the other. For a discussion of this issue, see Radder, 1994.

10. Nickles, 1988, 33 and 35–36. See also Nickles, 1989, esp. 309–313.

11. Gooding, 1990, 4–9, offers a discussion and classification of six different types or stages of reconstruction: cognitive, demonstrative, methodological, rhetorical, didactic, and philosophical. His book focuses on the first two types and provides a number of important insights into what goes on at these early stages of scientific inquiry. Yet, his discussion does not seem to capture the full significance of Nickles's approach. First, it still sticks to the Kuhnian dichotomy by characterizing cognitive and demonstrative reconstructions, in contrast to the other types, as "real time." Does this imply that, for instance, textbooks are written, used, or interpreted outside of "real time"? Second, on the basis of his analyses of the first two types, Gooding launches a number of strong criticisms of other, mainly philosophical, interpretations of science. On his own approach, however, these criticisms are either premature or an instance of the genetic fallacy. After all, the later stages of reconstruction might well endow the relevant scientific practices, processes, and products with a transformed meaning, which might (or might not) be adequately captured by the criticized interpretations.

12. Brown, 1988, 53. See also Quine, 1985; Goldman, 1986. With respect to scientific knowledge, see, e.g., Giere, 1985, 1988; Fuller, 1991.

13. Cf. also Stump, 1992. Note that my point here is *not* the "circularity" of the naturalist conception (discussed for instance by Brown, 1988, and by Giere, 1985), but rather its lack of a specific and positive content.

14. Just think of Husserl, who entitled one of his influential books *Philosophie als strenge Wissenschaft* (*Philosophy as a Rigorous Science*).

15. See Laudan, 1987, and Brown, 1988. In his 1988 book, Giere also takes this position.

16. Notably enough, the spectacular recent advances in this field are mainly due to increased "brute force" capacities and far less to improved "intelligence." This should (but does not always) lead to a considerably more modest assessment of the future potential of artificial intelligence projects in areas (such as science or jurisdiction) in which problems are much more complicated than in chess. After all, chess problems are relatively simple in that the allowed moves of the pieces are unambiguously fixed, while the possible outcomes of the game (win, lose, or draw) can be defined on the basis of a set of explicit rules.

17. Cf. the claim by Haugeland, 1985, 48, that chess *is* simply a formal system.

18. In Radder, 1988, ch. 2, I have shown this in detail for the case of Habermas's quasi-transcendental theories of truth and objectivity.

19. See also the illuminating discussion in Pels, 1990.

20. Cf. also Haraway, 1991, 193: "The split and contradictory self is the one who can interrogate positionings and be accountable, the one who can construct and join rational conversations and fantastic imaginings that change history."

REFERENCES

Abstracts of the 1988 Joint 4S/EASST Conference, 1988, *Science, Technology and Human Values*, 13, 83–206.
Bachelard, G., 1972, *Le Matérialisme Rationnel* (Paris: Presses Universitaires de France).
———, 1984, *The New Scientific Spirit* (Boston: Beacon Press).
Baird, D., 1988, "Five Theses on Instrumental Realism," in A. Fine and J. Leplin, eds., *PSA 1988*, Vol. I (East Lansing: Philosophy of Science Association), 165–173.
Barbour, I. G., 1980, *Technology, Environment, and Human Values* (New York: Praeger).
Barnes, B., 1981, "On the 'Hows' and 'Whys' of Cultural Change," *Social Studies of Science*, 11, 481–498.
Barnes, B., and Bloor, D., 1982, "Relativism, Rationalism and the Sociology of Knowledge," in M. Hollis and S. Lukes, eds., *Rationality and Relativism* (Oxford: Basil Blackwell), 21–47.
Beck, C. I., and Ulrich, T., 1993, "Biotechnology in the Food Industry," *Bio/Technology*, 11, 895–902.
Bhaskar, R., 1975, "Feyerabend and Bachelard: Two Philosophies of Science," *New Left Review*, 94, Nov.–Dec., 31–55.
———, 1978, *A Realist Theory of Science* (Hassocks: Harvester Press).
———, 1986, *Scientific Realism and Human Emancipation* (London: Verso).
Biesiot, W., 1983, "De kontroverses over de effekten van lage doses ioniserende straling," *Wetenschap en Samenleving*, nr. 4, 11–18.
Bijker, W. E., 1990, *The Social Construction of Technology* (Enschede: Doctoral dissertation, Universiteit Twente).
———, 1993, "Do Not Despair: There Is Life after Constructivism," *Science, Technology and Human Values*, 18, 113–138.
Bijker, W. E., Hughes, T. P., and Pinch, T., eds., 1987, *The Social Construction of Technological Systems* (Cambridge, Mass.: MIT Press).
Blokhuis, P., 1985, *Kennis en abstraktie* (Amsterdam: VU-uitgeverij).
Bogen, J., and Woodward, J., 1988, "Saving the Phenomena," *The Philosophical Review*, 97, 303–352.
Bohr, N., 1913, "On the Constitution of Atoms and Molecules, Part I," *Philosophical Magazine*, 26, 1–25.
———, 1918, "On the Quantum Theory of Line Spectra," in B. L. Van der Waerden, ed., *Sources of Quantum Mechanics* (Amsterdam: North-Holland Publishing Company, 1967), 95–136.
———, 1976, *Collected Works*, Vol. 3 (Amsterdam: North-Holland Publishing Company).

Borgmann, A., 1984, *Technology and the Character of Contemporary Life* (Chicago: University of Chicago Press).
Born, M., 1924, "Quantum Mechanics," in B. L. Van der Waerden, ed., *Sources of Quantum Mechanics* (Amsterdam: North-Holland Publishing Company, 1967), 181–198.
Boyd, R., 1979, "Metaphor and Theory Change: What Is 'Metaphor' a Metaphor for?" in A. Ortony, ed., *Metaphor and Thought* (Cambridge: Cambridge University Press), 356–408.
Broerse, J. E. W., 1990, *Country Case Study Pakistan* (Amsterdam: Report of the Department of Biology and Society, Vrije Universiteit).
——, 1992, *Rhizobia Inoculant Technology for the Improvement of Groundnut Production in the Communal Areas of Zimbabwe* (Amsterdam: Report of the Department of Biology and Society, Vrije Universiteit).
Broerse, J. E. W., and Wessels, H., 1989, "Towards a Dutch Policy on Biotechnology and Development Cooperation," *Trends in Biotechnology*, 7, nr. 1, 25–27.
Brouwer, H., Stokhof, E. M., and Bunders, J. F. G., eds., 1992, *Biotechnology and Farmer's Rights* (Amsterdam: VU University Press).
Brown, H. I., 1988, "Normative Epistemology and Naturalized Epistemology," *Inquiry*, 31, 53–78.
Brown, J. R., 1989, *The Rational and the Social* (London: Routledge).
Bukman, P., 1989, "The Government Role in Biotechnology and Development Cooperation," *Trends in Biotechnology*, 7, nr. 1, 27–31.
Bunders, J. F. G., ed., 1990, *Biotechnology for Small-Scale Farmers in Developing Countries* (Amsterdam: VU University Press).
Bunders, J. F. G., and Broerse, J. E. W., eds., 1991, *Appropriate Biotechnology in Small-Scale Agriculture* (Wallingford: C. A. B. International).
Bunders, J. F. G., Broerse, J. E. W., and Stolp, A., 1989, "Necessary, Robust and Supportable: The Requirements of Appropriate Biotechnology," *Trends in Biotechnology*, 7, nr. 1, 16–24.
Bunders, J. F. G., and Leydesdorff, L., 1987, "The Causes and Consequences of Collaborations between Scientists and Non-Scientific Groups," in S. S. Blume, J. F. G. Bunders, L. Leydesdorff, and R. Whitley, eds., *The Social Direction of the Public Sciences* (Dordrecht: Reidel), 331–347.
Bunders, J. F. G., Sarink, H., and De Bruin, J., 1989, "Seeking a Common Language," *Trends in Biotechnology*, 7, nr. 1, 5–7.
Bunders, J. F. G., Stolp, A., and Broerse, J. E. W., 1991, "An Interactive Bottom-Up Approach in Agricultural Research," in Bunders and Broerse, *Appropriate Biotechnology*, 71–109.
Bunge, M., 1970, "Problems Concerning Intertheory Relations," in P. Weingartner and G. Zecha, eds., *Induction, Physics and Ethics* (Dordrecht: Reidel), 285–325.
Callon, M., 1987, "Society in the Making: The Study of Technology as a Tool for Sociological Analysis," in Bijker, Hughes, and Pinch, *Social Construction*, 83–103.
Callon, M., and Latour, B., 1981, "Unscrewing the Big Leviathan: How Actors Macro-structure Reality and How Sociologists Help Them to Do So," in K. D.

Knorr-Cetina and A. V. Cicourel, eds., *Advances in Social Theory and Methodology. Toward an Integration of Micro- and Macro-sociologies* (Boston, Mass.: Routledge and Kegan Paul), 277–303.

———, 1992, "Don't Throw the Baby Out with the Bath School! A Reply to Collins and Yearley," in Pickering, *Science*, 343–368.

Callon, M., and Law, J., 1982, "On Interests and Their Transformation: Enrolment and Counter-Enrolment," *Social Studies of Science*, 12, 615–625.

Callon, M., Law, J., and Rip, A., 1986, "How to Study the Force of Science," in M. Callon, J. Law, and A. Rip, eds., *Mapping the Dynamics of Science and Technology* (London: MacMillan Press), 3–15.

Cartwright, N., 1983, *How the Laws of Physics Lie* (Oxford: Clarendon Press).

Chubin, D. E., and Restivo, S., 1983, "The 'Mooting' of Science Studies: Research Programmes and Science Policy," in Knorr-Cetina and Mulkay, *Science Observed*, 53–83.

Collins, H. M., 1975, "The Seven Sexes: A Study in the Sociology of a Phenomenon, or the Replication of Experiments in Physics," *Sociology*, 9, 205–224.

———, 1981, "Stages in the Empirical Programme of Relativism," *Social Studies of Science*, 11, 3–10.

———, 1983, "An Empirical Relativist Programme in the Sociology of Scientific Knowledge," in Knorr-Cetina and Mulkay, *Science Observed*, 85–113.

———, 1985, *Changing Order: Replication and Induction in Scientific Practice* (London: Sage).

———, 1990, *Artificial Experts* (Cambridge, Mass.: MIT Press).

Constant, E. W., 1984, "Communities and Hierarchies: Structure in the Practice of Science and Technology," in R. Laudan, ed., *The Nature of Technological Knowledge* (Dordrecht: Reidel), 27–46.

Crocker, D. A., 1991, "Toward Development Ethics," *World Development*, 19, 457–483.

Darrigol, O., 1986, "The Origin of Quantized Matter Waves," *Historical Studies in the Physical and Biological Sciences*, 16, 197–253.

Dear, P., 1985, "*Totius in Verba*: Rhetoric and Authority in the Early Royal Society," *Isis*, 76, 145–161.

De Bruin, J., and Bunders, J. F. G., 1987, *Evaluation of the Perspectives for Cooperation between Plant Biotechnologists and Environmental, Third World and Farmers' Organizations* (Amsterdam: Report of the Department of Biology and Society, Vrije Universiteit).

De Ruiter, W., 1992, *De evolutie van de laser* (Eindhoven: Doctoral dissertation, Technische Universiteit Eindhoven).

De Vries, G., 1991, "Ethische theorieën en de ontwikkeling van medische technologie," *Kennis en Methode*, 13, 278–294.

DGIS, 1989, *Biotechnology and Development Cooperation: Inventory of the Biotechnology Policy and Activities of a Number of Donor Countries and Organizations, UN Agencies, Development Banks, and CGIAR* (The Hague, the Netherlands: Report of the Netherlands Directorate General for International Cooperation).

———, 1991a, *Cassava and Biotechnology* (The Hague, the Netherlands: Report of the Netherlands Directorate General for International Cooperation).

―――, 1991b, *The Impact of Intellectual Property Protection in Biotechnology and Plant Breeding on Developing Countries* (The Hague, the Netherlands: Ministry of Foreign Affairs).

Dingler, H., 1952, *Ueber die Geschichte und das Wesen des Experimentes* (Munich: Eidos Verlag).

Downey, G. L., 1988, "Reproducing Cultural Identity in Negotiating Nuclear Power: The Union of Concerned Scientists and Emergency Core Cooling," *Social Studies of Science*, 18, 231–264.

Draaisma, D., 1989, "Voorbij het getal van Avogadro: de Benveniste-affaire," *Kennis en Methode*, 13, 84–107.

Dresden, M., 1987, *H. A. Kramers: Between Tradition and Revolution* (New York: Springer Verlag).

Duesing, J., 1992, "The Convention on Biological Diversity. Its Impact on Biotechnology Research," *Agro-Food-Industry Hi-Tech*, July–Aug., 19–23.

Fadner, W. L., 1985, "Theoretical Support for the Generalized Correspondence Principle," *American Journal of Physics*, 53, 829–838.

Feyerabend, P. K., 1962, "Explanation, Reduction and Empiricism," in H. Feigl and G. Maxwell, eds., *Minnesota Studies in the Philosophy of Science*, Vol. III (Minneapolis: University of Minnesota Press), 28–97.

―――, 1989, "Realism and the Historicity of Knowledge," *The Journal of Philosophy*, 86, 393–406.

Feynman, R. P., Leighton, R. B., and Sands, M., 1964, *The Feynman Lectures on Physics*, Vol. II (Reading, Mass.: Addison-Wesley Publishing Company).

Fine, A., 1986, *The Shaky Game* (Chicago: University of Chicago Press).

Forman, P., 1971, "Weimar Culture, Causality and Quantum Theory, 1918–1927: Adaptation by German Physicists and Mathematicians to a Hostile Environment," *Historical Studies in the Physical Sciences*, 3, 1–115.

Fraley, R., and Schell, J., 1993, "Plant Biotechnology—Editorial Overview," *Current Opinion Biotechnology*, 4, 133–134.

Franklin, A., 1986, *The Neglect of Experiment* (Cambridge: Cambridge University Press).

Fulkerson, J. F., 1991, "Understanding the Impacts of Biotechnology," *Phytopathology*, 81, 343–361.

Fuller, S., 1988, *Social Epistemology* (Bloomington: Indiana University Press).

―――, 1991, "Naturalized Epistemology Sublimated: Rapprochement without the Ruts," *Studies in History and Philosophy of Science*, 22, 277–293.

―――, 1993, *Philosophy of Science and Its Discontents*, 2nd edition (New York: Guilford Press).

Galison, P., 1987, *How Experiments End* (Chicago: University of Chicago Press).

Gault, R., 1991, "Uit de greep van de technologie," *Krisis*, nr. 45, 20–33.

Giere, R. N., 1985, "Philosophy of Science Naturalized," *Philosophy of Science*, 52, 331–356.

―――, 1988, *Explaining Science* (Chicago: University of Chicago Press).

Gleick, J., 1987, *Chaos* (New York: Penguin Books).

Goggin, M. L., ed., 1986, *Governing Science and Technology in a Democracy* (Knoxville: University of Tennessee Press).

Goldman, A., 1986, *Epistemology and Cognition* (Cambridge, Mass.: Harvard University Press).
Gooding, D., 1985, "'In Nature's School': Faraday as an Experimentalist," in D. Gooding and F. James, eds., *Faraday Rediscovered* (New York: Stockton Press), 105–135.
——, 1986, "How Do Scientists Reach Agreement about Novel Observations?" *Studies in History and Philosophy of Science*, 17, 205–230.
——, 1989, "'Magnetic curves' and the Magnetic Field: Experimentation and Representation in the History of a Theory," in Gooding, Pinch, and Schaffer, *Uses of Experiment*, 183–223.
——, 1990, *Experiment and the Making of Meaning* (Dordrecht: Kluwer).
Gooding, D., Pinch, T. J., and Schaffer, S., eds., 1989, *The Uses of Experiment* (Cambridge: Cambridge University Press).
Goodman, N., 1973, *Fact, Fiction and Forecast* (Indianapolis: Bobbs-Merrill).
Gotsch, N., and Rieder, P., 1989, "Future Importance of Biotechnology in Arable Farming," *Trends in Biotechnology*, 7, 29–34.
Groenewegen, P., 1985, "Wetenschappelijke macht in de praktijk," *Krisis*, nr. 18, 23–36.
Gutting, G., 1980, "Science as Discovery," *Revue Internationale de Philosophie*, 131–132, 26–48.
Habermas, J., 1970, *Toward a Rational Society* (Boston: Beacon Press).
——, 1973, "Wahrheitstheorien," in H. Fahrenbach, ed., *Wirklichkeit und Reflexion. Festschrift für W. Schulz* (Pfullingen: Neske), 211–265.
——, 1978, *Knowledge and Human Interests*, 2nd edition (London: Heinemann).
Hacking, I., 1983a, "Beyond Good and Evil," in J. Kendrew and J. H. Snelley, eds., *Priorities in Research* (Amsterdam: Excerpta Medica), 37–42.
——, 1983b, *Representing and Intervening* (Cambridge: Cambridge University Press).
——, 1986, "Culpable Ignorance of Interference Effects," in D. MacLean, ed., *Values at Risk* (Totowa: Rowman and Allanheld), 136–154.
——, 1988a, "On the Stability of the Laboratory Sciences," *The Journal of Philosophy*, 85, 507–514.
——, 1988b, "The Participant Irrealist at Large in the Laboratory," *British Journal for the Philosophy of Science*, 39, 277–294.
——, 1989a, "Extragalactic Reality: The Case of Gravitational Lensing," *Philosophy of Science*, 56, 555–581.
——, 1989b, "Philosophers of Experiment," in A. Fine and J. Leplin, eds., *PSA 1988*, Vol. II (East Lansing: Philosophy of Science Association), 147–156.
Hagendijk, R., 1990, "Structuration Theory, Constructivism, and Scientific Change," in S. E. Cozzens and T. F. Gieryn, eds., *Theories of Science in Society* (Bloomington: Indiana University Press), 43–66.
Hakfoort, C., 1986, *Optica in de eeuw van Euler* (Amsterdam: Rodopi).
Haraway, D., 1991, *Simians, Cyborgs, Women* (London: Routledge).
Harvey, B., 1981, "Plausibility and the Evaluation of Knowledge: A Case-Study of Experimental Quantum Mechanics," *Social Studies of Science*, 11, 95–130.

Hassoun, C. Q., and Kobe, D. H., 1989, "Synthesis of the Planck and Bohr Formulations of the Correspondence Principle," *American Journal of Physics*, 57, 658–662.
Haugeland, J., 1985, *Artificial Intelligence: The Very Idea* (Cambridge, Mass.: MIT Press).
Heilbron, J. L., and Kuhn, T. S., 1969, "The Genesis of the Bohr Atom," *Historical Studies in the Physical Sciences*, 1, 211–290.
Henderson, D. K., 1990, "On the Sociology of Science and the Continuing Importance of Epistemologically Couched Accounts," *Social Studies of Science*, 20, 113–148.
Hendry, J., 1982, "Bohr-Kramers-Slater: A Virtual Theory of Virtual Oscillators," *Centaurus*, 25, 189–211.
Hesse, M., 1963, *Models and Analogies in Science* (London: Sheed and Ward).
———, 1986, "Changing Concepts and Stable Order," *Social Studies of Science*, 16, 714–726.
Hillman, J. R., 1992, "Opportunities and Problems in Plant Biotechnology—An Overview," *Proceedings of the Royal Society of Edinborough*, 99B, 173–182.
Hones, M. J., 1990, "Reproducibility as a Methodological Imperative in Experimental Research," in A. Fine, M. Forbes, and L. Wessels, eds., *PSA 1990*, Vol. I (East Lansing: Philosophy of Science Association), 585–599.
Hookway, C., 1988, *Quine: Language, Experience and Reality* (Stanford: Stanford University Press).
Horstman, K., 1988, "Dilemma's van professionele geschiedsschrijving," *Kennis en Methode*, 12, 73–85.
Hoyningen-Huene, P., 1987, "Context of Discovery and Context of Justification," *Studies in History and Philosophy of Science*, 18, 501–515.
Hughes, T. P., 1987, "The Evolution of Large Technological Systems," in Bijker, Hughes, and Pinch, *Social Construction*, 51–82.
Ihde, D., 1991, *Instrumental Realism* (Bloomington: Indiana University Press).
Jammer, M., 1966, *The Conceptual Development of Quantum Mechanics* (New York: McGraw-Hill).
Janich, P., 1978, "Physics—Natural Science or Technology?" in W. Krohn, E. T. Layton, and P. Weingart, eds., *The Dynamics of Science and Technology* (Dordrecht: Reidel), 3–27.
Jonas, H., 1979, *Das Prinzip Verantwortung: Versuch einer Ethik für die technologische Zivilisation* (Frankfurt am Main: Insel Verlag).
Jones, J. L., 1992, "Genetic Engineering of Crops: Its Relevance to the Food Industry," *Trends in Food Science and Technology*, 31, 54–59.
Keller, E. F., 1985, *Reflections on Gender and Science* (New Haven: Yale University Press).
———, 1988, "Feminist Perspectives on Science Studies," *Science, Technology and Human Values*, 13, 235–249.
Keulartz, J., Kwa, C.-L., and Radder, H., 1985, "Scientific and Social Problems and Perspectives of Alternative Medicine," *Radical Philosophy*, nr. 41, 2–9.
Kirschenmann, P. P., 1985, "Neopositivism, Marxism and Idealization. Some Comments on Professor Nowak's Paper," *Studies in Soviet Thought*, 30, 219–235.

———, 1990, "Heuristical Strategies: Another Look at Idealization and Concretization," in J. Brzeziński, F. Coniglione, T. A. F. Kuipers, and L. Nowak, eds., *Idealization. Volume I: General Problems* (Amsterdam: Rodopi), 227–240.

Knorr-Cetina, K. D., 1983, "The Ethnographic Study of Scientific Work: Towards a Constructivist Interpretation of Science," in Knorr-Cetina and Mulkay, *Science Observed*, 115–140.

Knorr-Cetina, K. D., and Mulkay, M., eds., 1983, *Science Observed* (London: Sage).

Koertge, N., 1969, *A Study of Relations Between Scientific Theories: A Test of the General Correspondence Principle* (London: Doctoral thesis, University of London).

———, 1973, "Theory Change in Science," in G. Pearce and P. Maynard, eds., *Conceptual Change* (Dordrecht: Reidel), 167–198.

Koningsveld, H., 1973, *Empirical Laws, Regularity and Necessity* (Wageningen: Veenman).

Krajewski, W., 1977, *Correspondence Principle and Growth of Science* (Dordrecht: Reidel).

Kramers, H. A., 1919, "Intensities of Spectral Lines," in H. A. Kramers, *Collected Scientific Papers* (Amsterdam: North-Holland Publishing Company, 1956), 1–108.

Krige, J., 1978, "Popper's Epistemology and the Autonomy of Science," *Social Studies of Science*, 8, 287–307.

Krohn, W., and Weyer, J., 1989, "Gesellschaft als Labor," *Soziale Welt*, 40, 349–373.

Kuhn, T. S., 1970a, "Logic of Discovery or Psychology of Research?" in I. Lakatos and A. Musgrave, eds., *Criticism and the Growth of Knowledge* (Cambridge: Cambridge University Press), 1–23.

———, 1970b, *The Structure of Scientific Revolutions*, 2nd edition (Chicago: University of Chicago Press).

Kunneman, H., and Hullegie, W., 1989, *Een tripolair model van lichamelijkheid* (Amsterdam: Unpublished Manuscript).

Laird, F. N., 1993, "Participatory Analysis, Democracy, and Technological Decision Making," *Science, Technology and Human Values*, 18, 341–361.

Lakatos, I., 1970, "Falsification and the Methodology of Scientific Research Programmes," in I. Lakatos and A. Musgrave, eds., *Criticism and the Growth of Knowledge* (Cambridge: Cambridge University Press), 91–195.

Lamb, C. J., Ryals, J. A., Ward, E. R., and Dixon, R. A., 1992, "Emerging Strategies for Enhancing Crop Resistance to Microbial Pathogens," *Bio/Technology*, 10, 1436–1445.

Latour, B., 1983, "Give Me a Laboratory and I Will Raise the World," in Knorr-Cetina and Mulkay, *Science Observed*, 141–170.

———, 1987a, "Hoe 'De Heerser' te schrijven voor zowel machines als machinaties?" *Krisis*, nr. 26, 42–66.

———, 1987b, *Science in Action* (Milton Keynes: Open University Press).

———, 1988, "The Politics of Explanation: An Alternative," in Woolgar, *Knowledge and Reflexivity*, 155–176.

Latour, B., and Woolgar, S., 1979, *Laboratory Life* (London: Sage).
Laudan, L., 1980, "Why was the Logic of Discovery Abandoned?" in T. Nickles, ed., *Scientific Discovery, Logic and Rationality* (Dordrecht: Reidel), 173–183.
———, 1981, "A Confutation of Convergent Realism," *Philosophy of Science*, 48, 19–49.
———, 1987, "Progress or Rationality? The Prospects for Normative Naturalism," *American Philosophical Quarterly*, 24, 19–31.
Laudan, R., 1984, "Introduction," in R. Laudan, ed., *The Nature of Technological Knowledge* (Dordrecht: Reidel), 1–26.
———, 1993, "The 'New' History of Science: Implications for Philosophy of Science," in D. Hull, M. Forbes, and K. Okruhlik, eds., *PSA 1992*, Vol. II (East Lansing: Philosophy of Science Association), 476–481.
Laudan, R., Laudan, L., and Donovan, A., 1988, "Testing Theories of Scientific Change," in A. Donovan, L. Laudan, and R. Laudan, eds., *Scrutinizing Science* (Dordrecht: Kluwer), 3–44.
Law, J., 1984, "International Workshop on New Developments in the Social Studies of Technology," *4S Review*, 2, nr. 4, 9–13.
———, 1987, "Technology and Heterogeneous Engineering: The Case of Portuguese Expansion," in Bijker, Hughes, and Pinch, *Social Construction*, 111–134.
Levi, P., 1986, *The Periodic Table* (London: Sphere Books).
Levy, D., 1992, "Computer Chess with FIDE Approval?!?" *ICCA Journal*, 15, 150–151.
Liboff, R. L., 1984, "The Correspondence Principle," *Physics Today*, 37, 50–55.
Lindsey, K., 1992, "Genetic Manipulation of Crop Plants," *Journal of Biotechnology*, 26, 1–28.
Lynch, M., 1992, "Extending Wittgenstein: The Pivotal Move from Epistemology to the Sociology of Science," in Pickering, *Science*, 215–265.
Lynch, M., Livingston, E., and Garfinkel, H., 1983, "Temporal Order in Laboratory Work," in Knorr-Cetina and Mulkay, *Science Observed*, 205–238.
Lynch, W. T., and Fuhrman, E. R., 1991, "Recovering and Expanding the Normative: Marx and the New Sociology of Scientific Knowledge," *Science, Technology and Human Values*, 16, 233–248.
MacKenzie, D., 1986, "Science and Technology Studies and the Question of the Military," *Social Studies of Science*, 16, 361–371.
———, 1987, "Missile Accuracy: A Case Study in the Social Processes of Technological Change," in Bijker, Hughes, and Pinch, *Social Construction*, 195–222.
———, 1989, "From Kwajalein to Armageddon? Testing and the Social Construction of Missile Accuracy," in Gooding, Pinch, and Schaffer, *Uses of Experiment*, 409–435.
MacKenzie, D., and Spinardi, G., 1988a, "The Shaping of Nuclear Weapon System Technology: US Fleet Ballistic Missile Guidance and Navigation. I: From Polaris to Poseidon," *Social Studies of Science*, 18, 419–463.
———, 1988b, "The Shaping of Nuclear Weapon System Technology: US Fleet Ballistic Missile Guidance and Navigation. II: 'Going for Broke'—The Path to Trident II," *Social Studies of Science*, 18, 581–624.

MacKenzie, D., and Wajcman, J., eds., 1985, *The Social Shaping of Technology* (Milton Keynes: Open University Press).
Messiah, A., 1969, *Quantum Mechanics*, Vol. I (Amsterdam: North-Holland Publishing Company).
Meyer-Abich, K.-M., 1965, *Korrespondenz, Individualität und Komplementarität* (Wiesbaden: Franz Steiner Verlag).
Mol, A., 1989, "Wat je zegt, ben je dat zelf?" *Kennis en Methode*, 13, 257–262.
Morrison, M., 1990, "Theory, Intervention and Realism," *Synthese*, 82, 1–22.
Mushita, A., 1992, "Seed Improvement and Farmers' Rights in Zimbabwe," in Brouwer, Stokhof, and Bunders, *Biotechnology*, 31–35.
Nagel, E., 1961, *The Structure of Science* (London: Routledge and Kegan Paul).
Nelkin, D., ed., 1984, *Controversy: Politics of Technical Decisions*, 2nd edition (Beverly Hills: Sage).
Neurath, M., and Cohen, R. S., eds., 1973, *Otto Neurath. Empiricism and Sociology* (Dordrecht: Reidel).
Nickles, T., 1973, "Two Concepts of Intertheoretic Reduction," *The Journal of Philosophy*, 70, 181–201.
———, 1977, "Heuristics and Justification in Scientific Research," in Suppe, *Structure*, 571–589.
———, 1980, "Introductory Essay: Scientific Discovery and the Future of Philosophy of Science," in T. Nickles, ed., *Scientific Discovery, Logic and Rationality* (Dordrecht: Reidel), 1–59.
———, 1985, "Beyond Divorce: Current Status of the Discovery Debate," *Philosophy of Science*, 52, 177–206.
———, 1986, "Remarks on the Use of History as Evidence," *Synthese*, 69, 253–266.
———, 1988, "Reconstructing Science: Discovery and Experiment," in D. Batens and J. P. van Bendegem, eds., *Theory and Experiment* (Dordrecht: Reidel), 33–53.
———, 1989, "Justification and Experiment," in Gooding, Pinch, and Schaffer, *Uses of Experiment*, 299–333.
Noble, D., 1979, "Social Choice in Machine Design," in A. Zimbalist, ed., *Case Studies in the Labor Process* (New York: Monthly Review Press), 18–50.
Nunn, J., 1990, "Should Computers Participate in Chess Tournaments?" *ICCA Journal*, 13, 45–46.
Nussbaum, R. H., 1985, "Survivor Studies and Radiation Standards," *Bulletin of the Atomic Scientists*, 41, Aug., 68–71.
Park, J. L., 1968, "Quantum Theoretical Concepts of Measurement: Part I," *Philosophy of Science*, 35, 205–231.
Pels, D., 1990, "De 'natuurlijke saamhorigheid' van feiten en waarden," *Kennis en Methode*, 14, 14–43.
Perez, C., and Soete, L., 1988, "Catching up in Technology: Entry Barriers and Windows of Opportunity," in G. Dosi, C. Freeman, L. Soete, and J. Silverberg, eds., *Technical Change and Economic Theory* (London: Pinter Publishers), 458–479.
Perkins, J. H., 1982, *Insects, Experts and the Insecticide Crisis* (New York: Plenum Press).

Perrow, C., 1984, *Normal Accidents* (New York: Basic Books).
Pickering, A., 1981, "The Hunting of the Quark," *Isis*, 72, 216–236.
——, 1989, "Living in the Material World: On Realism and Experimental Practice," in Gooding, Pinch, and Schaffer, *Uses of Experiment*, 275–297.
——, 1990, "Knowledge, Practice and Mere Construction," *Social Studies of Science*, 20, 682–729.
——, ed., 1992, *Science as Practice and Culture* (Chicago: University of Chicago Press).
Pinch, T., 1985, "Towards an Analysis of Scientific Observation: The Externality and Evidential Significance of Observational Reports in Physics," *Social Studies of Science*, 15, 3–36.
Pinch, T. J., and Bijker, W. E., 1984, "The Social Construction of Facts and Artefacts: or How the Sociology of Science and the Sociology of Technology Might Benefit Each Other," *Social Studies of Science*, 14, 399–441.
——, 1986, "Science, Relativism and the New Sociology of Technology: Reply to Russell," *Social Studies of Science*, 16, 347–360.
Popper, K. R., 1965, *The Logic of Scientific Discovery* (New York: Harper and Row).
Post, H. R., 1971, "Correspondence, Invariance and Heuristics," *Studies in History and Philosophy of Science*, 2, 213–255.
Prasetyo, A. B., Timotius, K. H., Stouthamer, A. H., and Van Verseveld, H. W., 1991, *Socio-Economic Aspects of Indonesian Soy Sauce (Kecap) Production* (Universitas Kristen Satya Wacana, Indonesia: Report of the Faculty of Biology).
Price, D. Desolla, 1984, "The Science/Technology Relationship, the Craft of Experimental Science, and Policy for the Improvement of High Technology Innovation," *Research Policy*, 13, 3–20.
Price, D. K., 1979, "The Ethical Principles of Scientific Institutions," *Science, Technology and Human Values*, nr. 26, 46–60.
Putnam, H., 1975, *Mind, Language and Reality. Philosophical Papers*, Vol. 2 (Cambridge: Cambridge University Press).
——, 1978, *Meaning and the Moral Sciences* (Boston: Routledge and Kegan Paul).
Quine, W. V. O., 1985, "Epistemology Naturalized," in H. Kornblith, ed., *Naturalizing Epistemology* (Cambridge, Mass.: MIT Press), 15–29.
Radder, H., 1982, "Between Bohr's Atomic Theory and Heisenberg's Matrix Mechanics. A Study of the Role of the Dutch Physicist H. A. Kramers," *Janus*, 69, 223–252.
——, 1983, "Kramers and the Forman Theses," *History of Science*, 21, 165–182.
——, 1988, *The Material Realization of Science* (Assen: Van Gorcum). [Originally published as *De materiële realisering van wetenschap* (Amsterdam: VU-uitgeverij, 1984)].
——, 1989, "Rondom realisme," *Kennis en Methode*, 13, 295–314.
——, 1994, "Wetenschapsfilosofie en wetenschapsonderzoek: op weg naar een vruchtbare lat-relatie?" *Kennis en Methode*, 18, 157–168.
Radical Science Journal Collective, 1981, "Science, Technology, Medicine and the Socialist Movement," *Radical Science Journal*, nr. 11, 3–70.

Ravetz, J. R., 1973, *Scientific Knowledge and Its Social Problems* (Harmondsworth: Penguin Books).
Redhead, M. L. G., 1975, "Symmetry in Intertheory Relations," *Synthese*, 32, 77–112.
Reijnders, L., 1984, "Klopt de aanvaardbaar geachte dosis radio-aktiviteit nog wel?" *De Groene Amsterdammer*, 108, 7 November, 5.
Richards, E., and Schuster, J., 1989, "The Feminine Method as Myth and Accounting Resource: A Challenge to Gender Studies and Social Studies of Science," *Social Studies of Science*, 19, 697–720.
Richter, S., 1980, "Die 'Deutsche Physik,'" in H. Mehrtens and S. Richter, eds., *Naturwissenschaft, Technik und NS-Ideologie* (Frankfurt am Main: Suhrkamp), 116–141.
Ringwood, A. E., and Willis, P., 1984, "Stress Corrosion in a Borosilicate Glass Nuclear Wasteform," *Nature*, 311, 25 October, 735–737.
Röbbelen, G., 1990, "Mutation Breeding for Quality Improvement: A Case Study for Oilseed Crops," *Mutation Breeding Review*, 6, 1–44.
Rohrlich, F., 1988, "Pluralistic Ontology and Theory Reduction in the Physical Sciences," *British Journal for the Philosophy of Science*, 39, 295–312.
Rohrlich, F., and Hardin, L., 1983, "Established Theories," *Philosophy of Science*, 50, 603–617.
Rorty, R., 1979, *Philosophy and the Mirror of Nature* (Princeton: Princeton University Press).
Roth, P., and Barrett, R., 1990, "Deconstructing Quarks," *Social Studies of Science*, 20, 579–632.
Rouse, J., 1987, *Knowledge and Power* (Ithaca: Cornell University Press).
Russell, S., 1986, "The Social Construction of Artefacts: A Response to Pinch and Bijker," *Social Studies of Science*, 16, 331–346.
Salmon, W. C., 1984, *Scientific Explanation and the Causal Structure of the World* (Princeton: Princeton University Press).
Sass, H.-M., 1984, "Technik: Analyse, Bewertung, Beherrschung," *Philosophische Rundschau*, 31, 1–22.
Schäfer, W., ed., 1983, *Finalization in Science* (Dordrecht: Reidel).
Schaffer, S., 1989, "Glass Works: Newton's Prisms and the Uses of Experiment," in Gooding, Pinch, and Schaffer, *Uses of Experiment*, 67–104.
Schwartz Cowan, R., 1985, "The Industrial Revolution in the Home," in MacKenzie and Wajcman, *Social Shaping*, 181–201.
Scovel, G., 1992, "Genetical Engineering and Its Role in Conventional Breeding Programs," *Acta Horticulturae*, 307, 101–107.
Shapere, D., 1977, "Scientific Theories and Their Domains," in Suppe, *Structure*, 518–565.
Shapin, S., 1982, "History of Science and its Sociological Reconstructions," *History of Science*, 20, 157–211.
———, 1988, "The House of Experiment in Seventeenth-Century England," *Isis*, 79, 373–404.
———, 1989, "The Invisible Technician," *American Scientist*, 77, 554–563.
Shapin, S., and Schaffer, S., 1985, *Leviathan and the Air-Pump: Hobbes, Boyle, and the Experimental Life* (Princeton: Princeton University Press).

Slezak, P., 1989, "Scientific Discovery by Computer as Empirical Refutation of the Strong Programme," *Social Studies of Science*, 19, 563–600.

Somerville, C. R., 1993, "New Opportunities to Dissect and Manipulate Plant Processes," *Philosophical Transactions of the Royal Society*, 339, 199–206.

Sommerfeld, A., 1917, "Die Drudesche Dispersionstheorie vom Standpunkte des Bohrschen Modelles und die Konstitution von H_2, O_2 und N_2," *Annalen der Physik*, 53, 497–550.

Star, S. L., and Griesemer, J. R., 1989, "Institutional Ecology, 'Translations' and Boundary Objects: Amateurs and Professionals in Berkeley's Museum of Vertebrate Zoology, 1907–39," *Social Studies of Science*, 19, 387–420.

Stone, I. F., 1988, *The Trial of Socrates* (Boston: Little, Brown, and Company).

Stump, D., 1988, "The Role of Skill in Experimentation: Reading Ludwik Fleck's Study of the Wasserman Reaction as an Example of Ian Hacking's Experimental Realism," in A. Fine and J. Leplin, eds., *PSA 1988*, Vol. I (East Lansing: Philosophy of Science Association), 302–308.

——, 1992, "Naturalized Philosophy of Science with a Plurality of Methods," *Philosophy of Science*, 59, 456–460.

Stuurgroep Maatschappelijke Discussie Energiebeleid, 1983, *Het Tussenrapport* (Den Haag).

——, 1984, *Het Eindrapport* (Leiden: Stenfert Kroese).

Suppe, F., ed., 1977, *The Structure of Scientific Theories* (Urbana: University of Illinois Press).

Tetens, H., 1987, *Experimentelle Erfahrung* (Hamburg: Felix Meiner Verlag).

Van de Bulk, R. W., 1991, "Application of Cell and Tissue Culture and *In Vitro* Selection for Disease Breeding—A Review," *Euphytica*, 56, 285–296.

Van den Belt, H., 1991, "Habermas *versus* Rorty: Rationaliteit *versus* Radicaal Constructivisme," in B. Gremmen and S. Lijmbach, eds., *Toegepaste filosofie in de praktijk* (Wageningen: Wageningen Studies in Sociology 32, Agricultural University), 146–162.

Van Fraassen, B., 1980, *The Scientific Image* (Oxford: Clarendon Press).

Van Woudenberg, R., 1990, "Scepsis, zekere gronden en de methode van de 'reflexiviteit,'" *Tijdschrift voor Filosofie*, 52, 251–279.

Veldhuyzen van Zanten, J., 1992, "Contours of a Strategy," in Brouwer, Stokhof, and Bunders, *Biotechnology*, 85–89.

Verhagen, H., 1984, "Anti-kernenergiebeweging gereed voor de strijd," *De Volkskrant*, 63, 3 November, 17.

Walker, M. U., 1993, "Keeping Moral Space Open," *Hastings Center Report*, 23, 33–40.

Winner, L., 1986, "Building the Better Mousetrap," in L. Winner, *The Whale and the Reactor* (Chicago: University of Chicago Press), 61–84.

——, 1993, "Social Constructivism. Opening the Black Box and Finding It Empty," *Science as Culture*, 3, nr. 16, 427–452.

Wittgenstein, L., 1974, *Ueber Gewissheit/On Certainty* (Oxford: Basil Blackwell).

Woolgar, S., 1988a, *Science: The Very Idea* (Chichester and London: Ellis Horwood/Tavistock).

Woolgar, S., ed., 1988b, *Knowledge and Reflexivity* (London: Sage).

Woolgar, S., 1991, "The Very Idea of Social Epistemology: What Prospects for a Truly Radical 'Radically Naturalised Epistemology'?" *Inquiry*, 34, 377–389.

Woolgar, S., and Ashmore, M., 1988, "The Next Step: An Introduction to the Reflexive Project," in Woolgar, *Knowledge and Reflexivity*, 1–11.

Wynne, B., 1982, *Rationality and Ritual* (Chalfont St. Giles, Bucks.: The British Society for the History of Science).

———, 1983, "Redefining the Issues of Risk and Public Acceptance," *Futures*, 15, 13–32.

———, 1988, "Unruly Technology: Practical Rules, Impractical Discourses and Public Understanding," *Social Studies of Science*, 18, 147–167.

Zahar, E., 1976, "Why did Einstein's Programme Supersede Lorentz's?" in C. Howson, ed., *Method and Appraisal in the Physical Sciences* (Cambridge: Cambridge University Press), 211–276.

———, 1983, "Logic of Discovery or Psychology of Invention?" *British Journal for the Philosophy of Science*, 34, 243–261.

INDEX

abstraction, through replication, 35, 37–38, 83–86
actor-network theory, 102, 109–115, 198, 200
actors, in science and technology, 42–43, 94, 110–112, 123, 139–142, 197
appropriate technology, 137–138, 147–153, 166–167. *See also* biotechnology, appropriate agricultural
artificial intelligence. *See* chess, and computers
Ashmore, M., 99
atomic physics, 47, 53–57, 66–67
Avogadro's hypothesis, 23, 35

Bachelard, G., 86, 88, 195
Bacon, F., 134
Baigrie, B. S., ix
Baird, D., 195
Barbour, I. G., 152
Barnes, B., 95, 197
Barrett, R., 199
Beck, C. I., 201
Berlin, I., 198
Bernoulli, J., 47
Bhaskar, R., 45, 74, 86–88, 120, 190, 198, 200
Biesiot, W., 125
Bijker, W. E., 96, 107, 173–174, 197, 199–202
biotechnology: agricultural, 138–140, 153–166; appropriate agricultural, 143, 153–166; commercialization of, 164; and ethics, 137–138, 140–143

black box, 41–43, 102, 146, 192
Blokhuis, P., 85, 196
Bloor, D., 95
Bodewitz, H., x
Bogen, J., 12, 36–38, 195
Bohr, N., 47, 53–55, 57, 193
Bohr's atomic theory, 53–56, 66
Borgmann, A., 10, 42, 102, 146–147
Born, M., 53, 56, 66
Born's quantization rule, 56–57, 59, 65–66
Boyd, R., 46
Boyle, R.: air-pump experiments by, 15–16, 21–25, 27–29, 39–40, 43, 104, 196; gas law of, 49, 70
Broerse, J. E. W., 151, 153–154, 156, 159, 163, 165–166, 201
Brouwer, C., 197
Brouwer, H., 140, 164, 201
Brown, H. I., 179–180, 202–203
Brown, J. R., 191–192
Buchwald, J. Z., ix
Bukman, P., 163
Bunders, J. F. G., x, 140, 142, 144, 151–154, 156–157, 159, 164–166, 200–201
Bunge, M., 62

Callon, M., 97, 102, 109–113, 171–172, 200–201
Cartwright, N., 62, 86, 104, 190, 196, 200
Cassirer, E., 196
chess, and computers, 181–183, 203
Chubin, D. E., 196
Clausius, R., 47

219

closed systems: definition of, 119–123; in science and technology, 22, 86–87, 114, 123–128, 133–135, 199–201
Cohen, R. S., 176
Collins, H. M., 21, 23–24, 26–27, 30–35, 86, 95–96, 103, 108, 163, 190–192, 195–196, 198, 200
common language, 14, 17, 37, 80
conceptual discontinuity, 60–61, 75–76, 82. *See also* correspondence, conceptual and formal
Constant, E. W., 199, 201
constructivism, 88–89, 93–117, 171–174, 197–199, 202. *See also* social constructivism
control: of reality, 1–3; in science and technology, 22, 111, 114, 121–127, 133–134, 147, 161–162
correspondence: conceptual, 55–57, 59–60, 63–64, 193; formal, 55–63, 66, 69, 71, 76, 193; numerical, 53–63, 69, 71, 193
correspondence principle, in quantum theory, 47, 49–50, 53–59. *See also* generalized correspondence principle
Crick, F., 47
Crocker, D. A., 142

Darrigol, O., 194
De Broglie, L., 47
De Bruin, J., 142
De Ruiter, W., 33
De Vries, G., 150
Dear, P., 39, 43
delocalization, 29, 35–36
Dicke, R. H., 90
Dingler, H., 24, 189
Dirac, P. A. M., 47
discovery: scientific, 45, 71, 79, 87, 194; and justification, 52, 71–72, 176, 194
division of labor, 13–14, 17, 43, 146, 160, 202

Dixon, R. A., 201
Donovan, A., 189
Downey, G. L., 197
Draaisma, D., 191
Dresden, M., 64
Duesing, J., 200

effect thinking, 129–133
Ehrenfest theorem, 58, 64, 68
Einstein, A., 47, 62
empiricism, 91, 97, 171–174, 202
entomology, 126–127
epistemological positivism, 66–67
Epstein, P., 193
ethics: framework of, 137, 150, 166; recommendations from, 140–143, 150–151
experiment: boiling-point, 11–13, 17–18, 38; as craftwork, 22, 29, 44; material realization of, 11–22, 33, 37, 43, 84, 90, 109, 170, 189–190, 195, 201–202; theoretical description of, 11–14, 32, 36, 39–41, 76, 84–85, 90–91, 121–122, 125–128, 189–190. *See also* reproducibility
experimental technology, 40–43
experimenters' regress, 29–36, 191–192

Fadner, W. L., 46–48, 52–53, 57, 62, 65, 194
Faraday, M., 27–28, 35, 39, 104
farming systems, 157–160
Fehér, M., ix
feminist critiques, 32, 87, 95–96
Feyerabend, P. K., 60, 68, 195
Feynman, R. P., 61
Fine, A., 197
Forbes, M., ix
Forman, P., 66
Fraley, R., 201
Franklin, A., 191, 195
Fresnel, A. J., 47
Fuhrman, E. R., 196–197

Fulkerson, J. F., 200
Fuller, S., ix, 2, 113, 176–177, 189, 196, 202

Galison, P., 46, 191, 195, 197
Garfinkel, H., 15, 101, 171, 190, 202
Gault, R., x, 186
generalized correspondence principle: accounts of, 47–53; evaluation of, 59–64; heuristic scope of, 48, 51–52, 63–66
Giere, R. N., 174, 179–180, 190, 202–203
Gleick, J., 20
Goggin, M. L., 151
Goldenberg, M., 90
Goldman, A., 202
Gooding, D., 27, 35, 39–40, 46, 190–191, 195, 201–202
Goodman, N., 194
Gotsch, N., 200
Griesemer, J. R., 197
Groenewegen, P., x, 126
Gutting, G., 71

Habermas, J., 24, 86, 129, 176, 181, 189–190, 192, 200, 203
Hacking, I., 12, 21, 38, 46, 86, 91–92, 123, 131, 150, 190–191, 195–196
Hagendijk, R., 197
Hakfoort, C., 193
Hanson, N. R., 60
Haraway, D., 203
Hardin, L., 65
Harré, R., 86
Harvey, B., 23
Hassoun, C. Q., 58
Haugeland, J., 203
Heidegger, M., 196
Heilbron, J. L., 53
Heisenberg, W., 47, 53, 56–57, 66, 86
Henderson, D. K., 191
Hendry, J., 193

Hesse, M., 32, 35, 46
heuristic rules, 45–47, 50–52, 56–57, 64, 67, 71
Hilbert space, 59, 69
Hill, H., 90
Hillman, J. R., 200
Hobbes, T., 21, 40, 196
Hones, M. J., 27, 191
Hookway, C., 179
Horstman, K., 196
Hoyningen-Huene, P., 194
Hughes, T. P., 102, 107, 113, 198, 201
Hull, D., ix
Hullegie, W., 190
Husserl, E., 203

Illich, I., 134
input production systems, for agricultural biotechnology, 160–162
interactive bottom-up approach, 165–166

Jammer, M., 47, 56
Janich, P., 189
Jonas, H., 140
Jones, J. L., 200
justification, of scientific knowledge, 4, 49–50, 71–72, 108, 175–176, 184, 192, 194–195. *See also* discovery, and justification

Kant, I., 85, 117, 142
Keller, E. F., 87, 96, 103, 196–197
Kepler's laws, 51
Keulartz, J., x, 200
Kirschenmann, P. P., x, 50, 194
Knorr-Cetina, K. D., 101, 194–195, 198
knowledge: and power, 1–2, 87–88, 127, 133–134, 181, 200; transformation of, 45–46, 72, 175
Kobe, D. H., 58

Koertge, N., 60, 194
Koningsveld, H., 196
Krajewski, W., 46, 48–50, 52, 57, 61, 65, 70, 193
Kramers, H. A., 53–57, 64, 66, 193
Krige, J., 176
Kuhn, T. S., 53, 60, 68, 113, 178, 193, 195, 199
Kuhn loss, 48, 63, 68–69
Kunneman, H., 190
Kwa, C.-L., x, 200

Laird, F. N., 151
Lakatos, I., 46, 50, 52
Lamb, C. J., 200
Latour, B., 10, 27, 41–43, 74, 86–89, 95, 101–102, 109–113, 117, 123, 128, 134, 145, 172, 192, 194–195, 197–200
Laudan, L., 60, 70–71, 180, 189, 194, 203
Laudan, R., 189, 199, 202
Law, J., 97, 109–112, 123, 171, 198, 200–201
laws, of nature, 49, 51, 61, 70, 86–87, 104, 196
layperson, 13–17, 21–22, 34, 43, 170
Leighton, R. B., 61
Levi, P., 199
Levy, D., 183
Leydesdorff, L., 165
Liboff, R. L., 58
Lindsey, K., 201
Livingston, E., 15, 101, 171, 190, 202
localism, strict, 88–89
locality, 27, 101–105, 172
Locke, J., 85, 196
Lynch, M., 15, 101, 171, 190, 202
Lynch, W. T., 196–197

Machiavelli, N., 97
MacKenzie, D., 96, 102, 104, 109, 127–128, 198–199
material realization. *See* experiment
Maxwell, J. C., 47

Mepham, T. B., x
Messiah, A., 58–59, 69
metaphilosophy, 7–8, 169–170
Meyer-Abich, K.-M., 47
Mol, A., x, 100
Morpurgo, G., 25
Morrison, M., 11, 190, 196
Mulkay, M., 194–195, 198
Mushita, A., 164

Nagel, E., 68, 202
natural proximity, of facts and norms, 115
naturalism, 4, 177–181, 183, 202. *See also* normative naturalism
Nelkin, D., 129
Neurath, M., 176
Neurath, O., 176
Nickles, T., 36, 68–69, 71, 175, 189, 197, 202
Noble, D., 96, 200
nonlocal meaning, 2–3, 78
nonlocal patterns, 2–4, 29, 45–46, 102–105, 145, 169–173, 194
norm of reproducibility, 27–28, 103
normative naturalism, 180–181, 183
normative reflexion, 93–94, 100–101, 115–117
normativity, and constructivism, 93–117, 128
nuclear energy: technology of, 124–126, 149; and technology policy, 129–133
Nunn, J., 183
Nussbaum, R. H., 125

observation, 36–37, 89–92, 177
Okruhlik, K., ix
ontology, of potentialities and realizations, 79–82, 86–87, 89, 92, 171
Oudshoorn, N., 197
ought-implies-can principle, 142, 152
ozone layer, hole in, 107–108, 110, 186, 198

Park, J. L., 194
Pasteur, L., 110, 123, 128
Pauli, W., 59, 66
Pekelharing, P., x
Pels, D., 115, 196, 203
perception, 90–91
Perez, C., 163
Perkins, J. H., 126–127
Perrin, J., 23
Perrow, C., 114–115, 117
phenomena, and data, 36–38, 195
philosophy, normative, 9–10, 175–183, 194–195
philosophy, reflexive, 183–187. See also normative reflexion
philosophy, theoretical, 9, 19–20, 135, 170–171, 174–175
Pickering, A., 12, 25, 122–123, 195
Pinch, T., 90–91, 96, 107, 196, 199–201
Planck, M., 47
Popper, K. R., 24, 52, 86, 129, 176, 178, 190, 196
Post, H. R., 46, 48–49, 52, 57, 60–65, 68–69
postmodernism, 3–4, 197
potentiality. See ontology
Potter, E., 197
Prasetyo, A. B., 155
Price, D. Desolla, 46
Price, D. K., 140
Putnam, H., 70, 195

Quine, W. V. O., 177, 179, 202

rationality, instrumental, 133, 180–183, 186
Ravetz, J. R., 190
realism: Bachelardian challenge to, 74, 77–79, 85–86; convergent, 69–70, 76; Kuhnian challenge to, 73–77, 92; referential, 71, 75–76, 81–82, 108–109, 195–196; transcendental, 85–87, 198

realization: historical contingency of, 79–81, 83; of science and technology, 1–3, 76–83, 143–144, 147–149, 181–183, 195. See also biotechnology, appropriate agricultural
reconstruction, in science, 36, 175, 197, 202–203
Redhead, M. L. G., 46
reduction: of contingency, 1–3; of theories, 68–69
reflexivity: constitutive, 97–101, 184–185; differentially situated, 185–187; foundational and skeptical, 184–185. See also normative reflexion
Reijnders, L., 125
relativism, 75–76, 95, 100, 106–109, 197–198
replicability, of the experimental result, 18–19, 22–24, 28–31, 35–38, 82–86, 89, 196
reproducibility: of the material realization, 16–19, 21–22, 32–33, 41–43, 76, 80–81, 109, 190–191; of the theoretical description, 17–19, 24–26, 81–82, 121, 199; and stability, 28–29, 32–33, 36, 103; types and ranges of, 16–19, 28–29, 80–83
reproduction, in experimental practice, 20–26, 32–36, 80–83, 88–89, 191
Restivo, S., 196
Richards, E., 103
Richter, S., 190
Rieder, P., 200
Ringwood, A. E., 124
Rip, A., x, 97, 171
risk, of technology, 125–126, 131–133, 149–150, 157
Röbbelen, G., 200
Rohrlich, F., 65, 68–69
Rorty, R., 99
Roth, P., 199
Rouse, J., 22, 27, 87, 102, 104, 145, 190–191, 196–197, 200

Royal Society, 21, 29, 39–40, 43, 192
Russell, S., 196, 202
Rutherford, E., 193
Ryals, J. A., 200

Salmon, W. C., 190
Sands, M., 61
Sarink, H., 142
Sass, H.-M., 199
Schäfer, W., 102
Schaffer, S., 15–16, 21, 23–24, 27, 29, 39, 102, 190–191, 195–196
Schell, J., 201
Schrecker, F., 15
Schuster, J., 103
Schwartz Cowan, R., 95
science-technology relationship, 4, 40–41, 123–128, 145–146, 162–163, 192, 199
Scovel, G., 201
Shapere, D., 46
Shapin, S., 15–16, 21–24, 27, 29, 39–40, 43, 102, 190–192, 194–196
skill, in experimental science, 14, 17, 21–22, 30–36, 192
Slezak, P., 199
social constructivism, 30, 96, 106–107, 110
social critique, 93, 176
social legitimation of science, 38–41, 43
Soete, L., 163
Somerville, C. R., 201
Sommerfeld, A., 53–54, 193
Spinardi, G., 104
standardization, 21–22, 25, 29, 35–36, 102, 104, 146, 191
Star, S. L., 197
Stokhof, E. M., 140, 164, 201
Stolp, A., 151, 156, 165, 201
Stone, I. F., 201
Stouthamer, A. H., 155
Stump, D., 190, 202
Suppe, F., 202

technological culture, 147, 186
technology: device paradigm of, 42–43, 102, 146; key features of, 143–148, 201; science-based, 42, 145–146, 162–163
technology assessment, 150, 157
technology policy, 87, 107, 113–115, 129–130, 133
technoscience, 4, 42, 112, 185
technoscientific success, 112–115
Tetens, H., 189
Timotius, K. H., 155
Tucker, G. A., x

Ulrich, T., 201
universals, 83–84, 196

Van de Bulk, R. W., 200
Van den Belt, H., 187
Van den Wijngaard, M. A., 197
Van der Waals, J. D., gas law of, 49, 70
Van Fraassen, B., 91, 172, 198, 202
Van Lieshout, P., x
Van Putten, T., x
Van Verseveld, H. W., 155
Van Woudenberg, R., 184
Veldhuyzen van Zanten, J., 141
Verhagen, H., 130

Wajcman, J., 102
Walker, M. U., 166
Ward, E. R., 200
Watson, J., 47
Weber, J., gravity-wave experiment by, 27
Weber, M., 114
Wessels, H., 163
Whitehead, A. N., 86
Willis, P., 124
Winner, L., 152–153, 202
Wiseman, J., x
Wittgenstein, L., 100

Woodward, J., 12, 36–38, 195
Woolgar, S., 27, 74, 86, 88, 98–101, 103, 106–107, 184, 194–195, 197–198, 202
world, in and about the, 2, 8, 75, 169, 187
Wynne, B., 103, 129, 197, 199–200

yam tissue culture project, 153–155, 158–163, 165
Young, T., 47

Zahar, E., 46, 48, 50–52, 57, 60–61, 63–67, 194